Identity, Culture, and the Science Performance

Volume 2

Performance and Science: Interdisciplinary Dialogues explores the interactions between science and performance, providing readers with a unique guide to current practices and research in this fast-expanding field. Through shared themes and case studies, the series offers rigorous vocabularies and methods for empirical studies of performance, with each volume involving collaboration between performance scholars, practitioners and scientists. The series encompasses the modalities of performance to include drama, dance and music.

SERIES EDITORS

John Lutterbie
Chair of the Department of Art and of Theatre Arts at Stony Brook University, USA

Nicola Shaughnessy
Professor of Performance at the University of Kent, UK

IN THE SAME SERIES

Affective Performance and Cognitive Science: Body, Brain and Being
Edited by Nicola Shaughnessy
ISBN 978-1-4081-8398-4

An Introduction to Theatre, Performance and the Cognitive Sciences
John Lutterbie
ISBN 978-1-4742-5704-6

Collaborative Embodied Performance: Ecologies of Skill
Edited by Kath Bicknell and John Sutton
ISBN 978-1-3501-9769-5

Identity, Culture, and the Science Performance, Volume 1: From the Lab to the Streets
Edited by Vivian Appler and Meredith Conti
ISBN 978-1-3502-3406-2

Performance and the Medical Body
Edited by Alex Mermikides and Gianna Bouchard
ISBN 978-1-4725-7078-9

Performance, Medicine and the Human
Alex Mermikides
ISBN 978-1-3500-2215-7

Performing Psychologies: Imagination, Creativity and Dramas of the Mind
Edited by Nicola Shaughnessy and Philip Barnard
ISBN 978-1-4742-6085-5

Performing Specimens: Contemporary Performance and Biomedical Display
Gianna Bouchard
ISBN 978-1-3500-3567-6

Performing the Remembered Present: The Cognition of Memory in Dance, Theatre and Music
Edited by Pil Hansen with Bettina Bläsing
ISBN 978-1-4742-8471-4

Theatre and Cognitive Neuroscience
Edited by Clelia Falletti, Gabriele Sofia and Victor Jacono
ISBN 978-1-4725-8478-6

Theatre, Performance and Cognition: Languages, Bodies and Ecologies
Edited by Rhonda Blair and Amy Cook
ISBN 978-1-4725-9179-1

Identity, Culture, and the Science Performance

Volume 2

From the Curious to the Quantum

Edited by
Meredith Conti and Vivian Appler

methuen | drama
LONDON • NEW YORK • OXFORD • NEW DELHI • SYDNEY

METHUEN DRAMA

Bloomsbury Publishing Plc, 50 Bedford Square, London, WC1B 3DP, UK
Bloomsbury Publishing Inc, 1385 Broadway, New York, NY 10018, USA
Bloomsbury Publishing Ireland, 29 Earlsfort Terrace, Dublin 2, D02 AY28, Ireland

BLOOMSBURY, METHUEN DRAMA and the Methuen Drama logo are trademarks of Bloomsbury Publishing Plc

First published in Great Britain 2024
This paperback edition published in 2025

Copyright © Meredith Conti, Vivian Appler, and contributors, 2024

Meredith Conti and Vivian Appler have asserted their right under the Copyright, Designs and Patents Act, 1988, to be identified as Editors of this work.

For legal purposes the Acknowledgments on p. xi constitute an extension of this copyright page.

Series design by Louise Dugdale

Cover image: Artists in the costumes of a molecular crystal lattice during the festival of science, art and technology "Polytech" in Gorky Park in Moscow, Russia (© Nikolay Vinokurov / Alamy Stock Photo)

All rights reserved. No part of this publication may be: i) reproduced or transmitted in any form, electronic or mechanical, including photocopying, recording or by means of any information storage or retrieval system without prior permission in writing from the publishers; or ii) used or reproduced in any way for the training, development or operation of artificial intelligence (AI) technologies, including generative AI technologies. The rights holders expressly reserve this publication from the text and data mining exception as per Article 4(3) of the Digital Single Market Directive (EU) 2019/790.

Bloomsbury Publishing Plc does not have any control over, or responsibility for, any third-party websites referred to or in this book. All internet addresses given in this book were correct at the time of going to press. The author and publisher regret any inconvenience caused if addresses have changed or sites have ceased to exist, but can accept no responsibility for any such changes.

A catalogue record for this book is available from the British Library.

A catalog record for this book is available from the Library of Congress.

ISBN: HB: 978-1-3502-3426-0
PB: 978-1-3502-3430-7
ePDF: 978-1-3502-3428-4
eBook: 978-1-3502-3427-7

Series: Performance and Science: Interdisciplinary Dialogues

Typeset by Deanta Global Publishing Services, Chennai, India

For product safety related questions contact productsafety@bloomsbury.com.

To find out more about our authors and books visit www.bloomsbury.com and sign up for our newsletters.

For our dads,
Michael and David

CONTENTS

List of Illustrations x
Acknowledgments xi

Introduction: Entangled Domains: Revealing the Productive Tensions of Art and Science *Meredith Conti and Vivian Appler* 1

PART ONE Performing Human and More-Than-Human Relationships 11

Performance, Climate, and Colonization Roundtable *Claudia Barnett, Chantal Bilodeau, David Geary, Kirsten Lindquist, and Kim TallBear* 13

Creative Interlude: Critical Polyamorist 100s *Kim TallBear* 34

1 Staging the Mad Past: Performance, Criticism, and Historiography in Steppenwolf Theatre Company's *One Flew Over the Cuckoo's Nest* *Alexis Riley* 36

2 *Through Fish Eyes*: Raising Awareness of Ocean Degradation through Performance *Kasi V. Aysola and Madhvi J. Venkatesh* 55

3 Laboring the Medical: Female Bodies for Sale on the Contemporary Stage *Gianna Bouchard* 75

Creative Interlude: Please Let Me Shoot You: a monologue *Claudia Barnett* 93

PART TWO Challenging Traditions through the Science Performance 99

4 Spooky('s) Action at a Distance: Remixing Science and Performance in the Planetarium Show *Mike Vanden Heuvel* 101

5 Using Short Digital Films to Counter Stereotypes about Scientists of Color and from Marginalized Backgrounds *Mónica I. Feliú-Mójer* 119

6 Celestial Politics: Performance and the Cosmic Underclass *Felipe Cervera* 138

Creative Interlude: Mother *Chantal Bilodeau* 153

PART THREE Revising the Art-Science Repertoire 157

7 Performing and Negotiating Imperialism: Science, Agriculture, and Food in Puerto Rico *Teófilo Espada-Brignoni* 159

8 Identity Crisis in Interwar Germany: Brecht's *Leben des Galilei* and the Crisis of Science *Derek Gingrich* 179

9 "Let Science and Art Have at It": The Living Newspapers Perform Science to Promote Depression-Era Theatre/Squonk Performs Theatre to Promote Trump-Era Science *Emily B. Klein* 200

Creative Interlude: From *Variation for Three Voices on a Letter to Nature* *Diane Stubbings* 222

The Catastrophist Artists' Roundtable
   William DeMeritt, Martine Kei Green-Rogers,
   and Lauren Gunderson 228

Creative Interlude: From *The Catastrophist*
   Lauren Gunderson 244

Selected Bibliography 247
List of Contributors 254
Index 261

ILLUSTRATIONS

Figures

1. "A Recipe for Seasoned Sex" 16
2. Mother whale feeding her calf 61
3. Coral reef 63
4. Ocean brought to life in anger 64
5. The Federal Theatre Project performs *Spirochete* in Seattle, 1939 207
6. Squonk performs *Hand to Hand* at PPG Plaza in Pittsburgh, Pennsylvania, in 2019 213

Table

1. *One Flew Over the Cuckoo's Nest*, a Timeline of Adaptation 40

ACKNOWLEDGMENTS

The publication of a two-volume collection of essays, interviews, and creative works is, by necessity, a group undertaking. We are so grateful to the contributors to this volume, without whose creativity, curiosity, intrepid attention to detail, and multiple communications this project would not have come to fruition. Our thanks go out to the editorial team at Methuen Drama, especially to Mark Dudgeon for overseeing the process and to Ella Wilson for her consistent contact and support. We remain grateful to series editors Nicola Shaughnessy and John Lutterbie for seeing this project through to its completion.

Like *Volume 1: From the Lab to the Streets*, many of the pieces published in this volume reflect ideas generated at the American Society for Theatre Research (ASTR) in 2016, 2017, and 2018. Although this collection is being published in different times, our early conversations and collaborations at ASTR remain central to the ideas circulating in this book series. We thank all of our past working group members for their generosity of time and curiosity of spirit; these volumes have been all the better for each of them.

Many thanks to our colleagues who make up the faculty and staff of our home institutions, the University at Buffalo, SUNY (UB), and the University of Georgia (UGA). Vivian appreciates the time taken by her multidisciplinary friends and colleagues Melissa Hughes, Sarah Hatteberg, and Shondrika Moss-Bouldin, who provided valuable feedback and writing support during this process. We would like to thank Raquell Holmes for our early discussions about this collection. Our gratitude also extends to Ian Downes for their transcript work and to the indexer for this volume, Shelby Brewster. Vivian would also like to thank UGA's Franklin College of Arts and Sciences and Office of the Provost for supporting various aspects of this collaborative interdisciplinary project.

Scene 3, 'Beeswax' © Lauren Gunderson, 2021, *The Catastrophist*, Methuen Drama, an imprint of Bloomsbury Publishing Plc. The editors and publisher gratefully acknowledge the permission granted to reproduce the copyright material in this book.

Introduction

Entangled Domains: Revealing the Productive Tensions of Art and Science

Meredith Conti and Vivian Appler

> *Storm still.*
> KING LEAR, ACT 3, SCENE 1[1]

The Bristol Old Vic's roof roared and trembled with booms of thunder. A newly restored technology of the theatre's eighteenth-century past, the thunder run had been pressed back into service for the company's 2016 remounting of *The Tragedy of King Lear*, inviting contemporary audiences to hear just what their Georgian counterparts heard as the storm battered the heath and its inhabitants. Last used in production in 1942, the Old Vic's thunder run remains one of only three in use in England today.[2] To simulate the sounds of thunder, stage technicians roll heavy balls down a long series of wooden troughs threaded through the theatre's ceiling.[3] A thunder run was, in effect, a surround-sound system for the early modern age, one that enveloped audiences in the reverberations of storms or war while remaining hidden from view, secreted within the theatre's very architecture.[4]

The expressive power of thunder runs was not limited to mimicking meteorological phenomena. As an "acousmatic play . . . replete with rumbling noises and blasting winds as well as with ballads and snatches of songs," *King Lear*'s tempestuous soundscape makes Lear's—and his kingdom's—undoing echo through the heavens, reinforcing the connectedness of the three astrological levels (family, state, cosmos).[5] Thunder run technology realized and amplified the playwright's provocation that nature be an active scene partner throughout the play, with Lear, the Earl of Gloucester, and Edmund variously railing against its destructive capacity, contemplating its influence on human behavior, marveling in (or fearing) its presumed powers of prognostication, or dismissing any causal relationship between natural forces and humanity.[6] Of course, Shakespeare does harness nature's power to convey the inner turmoil of the king. Lear indicates the onset of his distress in Act Two when he acknowledges, "we are not ourselves/ When nature, being oppress'd, commands the mind/ To suffer the body" (2.4.107–9); the storm is introduced in stage directions only after he is cast out by daughters Regan and Goneril at the end of the act. As the king attempts to master his emotions, the tempest materializes:

> No, I'll not weep:
> I have full cause of weeping but this heart
>   *Storm and tempest.*
> Shall break into a hundred thousand flaws,
> Or ere I'll weep, O fool, I shall go mad! (2.4.283–6)

Summoned into radical action in Act Three, nature strikes, and the stage directions comprising our epigraph repeat multiple times throughout the ensuing scenes.

The storm on the heath is both a natural weather occurrence and a metaphorical articulation of the king's deepening rage, desperation, and senility; the thunder run operated beyond the proscenium to externalize for spectators Lear's psychic turmoil and mental decline. Because the Georgian run's wooden gutters flowed through the Old Vic's ceiling, shunting balls over the heads of theatregoers, its effect was simultaneously magical and material. Whereas London's often ominous skies were visible to the Globe's groundlings during early public productions of *King Lear*, encouraging audiences to connect real and theatrical events of "meteorological violence," the Old Vic's thunder run loomed overhead, masked behind the auditorium's ornate ceiling.[7] The theatre's restored thunder run is now the central draw of "The Thunder Run Experience," a forty-five-minute public tour in which visitors are invited to "hear the theatrical thunder roar, just as it did 250 years ago."[8]

The Old Vic's *King Lear* revival integrated an Elizabethan tragedy, Georgian stage machinery, and twenty-first-century storytelling methods and sensibilities. Furthermore, whether or not the audience consciously processed the 2016 production as such, *King Lear* is a robustly science-integrative play, one that also gestures toward identity issues pertinent to contemporary audiences. Through its multisensory presentation of the king's "madness,"[9] the Old Vic production alluded to or interfaced with a number of STEM disciplines (psychology; natural philosophy; climatology and ecology; and engineering and technology), all the while demonstrating that the natural world often expresses itself in ways that are invisible, ephemeral, or inscrutable. Historic thunder runs exemplify theatre's age-old capacity to simultaneously conceal illusory technologies and reveal aspects of the human experience that are obscured by society's mores and cultural conventions.

Why *King Lear*? Why introduce an interdisciplinary volume on science, performance, and identity—one that delights in the field-shifting potentialities of radical inclusion, decolonizing processes, and diverse representation—using a stalwart from the Shakespearean canon? We begin with the Old Vic's *King Lear* revival precisely because both the play and its historical technics challenge a number of false binaries that this volume attempts to controvert: art and science, then and now, nature and society, mind and body, entertainment and education. Moreover, *Lear*'s centuries-spanning production history attests to theatre's importance as a forum for scientific inquiry among its makers and its publics. This is "[p]art of theatre's task . . . to make the invisible visible," as performance scholar Andrew Sofer puts it.[10] The staging of science questions within theatrical spaces can lead to complicated, nuanced discussions of human ontology and identity; indeed, *Lear* scholars continue to diagnose the king's physical and mental ailments, presenting Lear as an illness role informed by early modern understandings of senility and madness.[11] Such interdisciplinary conversations can push beyond the indexical—pointing not only to "something else connected to it not by resemblance . . . but by cause and effect" and suggesting what Sofer describes as gravitational relationship—the connection between physical objects and, construed within Sofer's metaphor, theatricalized ideas.[12] As *Lear* and its thunder runs demonstrate, when theatrical performances incorporate science (be it through content, methodologies, or technologies), the presumed opacity of scientific disciplines and discourses softens and allows the public to apprehend that which ivory towers and laboratory walls often hide from view.

Well into the quantum age, humans continue to encounter invisible forces that impact our lives in unpredictable, sometimes imperceptible ways. It is no surprise, then, that artists and scientists alike have regularly found inspiration in the enigmatic forces at work in the universe. Volume 2 of *Identity, Culture, and the Science Performance: From the Curious to the Quantum* explores how science performances—events that bridge theatrical and scientific fields— help us identify, assess, and make sense of hidden phenomena and mysterious matter(s) that shape human experience. Furthermore, *From the Curious to the Quantum* demonstrates that science and art are equally generative in humanity's quest to uncover the universe's secrets. Contributors to this volume analyze theatrical stagings of invisible or obscured scientific processes—private clinical trials, the hunt for microscopic viruses, emotions of marine animals—and by doing so affirm theatre's unique facility for imagining and producing phenomena not easily observed by nonexperts. This book therefore diverges from and complements the series' first volume, which prioritizes public, hyper-visible performances of science, from K-12 educational outreach programs and immersive museum exhibits to Victorian surgical demonstrations and plays on anatomizing and global organ harvesting.

Like Volume 1, Volume 2 is concerned with the sociocultural intersections of science, performance, and identity. *From the Curious to the Quantum* recognizes that both the sciences and the performing arts have historically privileged the white heteropatriarchy and marginalized, exploited, or discredited those who did not or could not conform to its ideologies. As a number of our contributors propose, the science performance can partly redress such histories by decentering whiteness; interrogating institutionalized racism, ableism, sexism, classism, and homo/transphobia; and centering diverse bodies, voices, ideas, and methods. Employing frameworks and methodologies drawn from the arts, humanities, and sciences,

From the Curious to the Quantum's featured scholars and artists continually foreground considerations of identity (as both a personal and collective category) and culture as they evaluate an array of science-integrative performances and interdisciplinary topics.

From the Curious to the Quantum is comprised of three parts, each boasting a selection of critical essays, artists' roundtables, and creative works. In Part One: Performing Human and More-Than-Human Relationships, contributors analyze theatre and dance performances that both imagine humans capable of caring for each other and the planet and confront the imposing institutional and societal barriers that often inhibit empathetic relationships. We open the collection with a roundtable on performance, climate, and colonialism in which the panelists, many of whom have generously contributed creative works to this collection, discuss performance's potential to intervene in present-day colonialist structures and practices. The roundtable's participants together put forth performance, and especially performance that reframes the human and nonhuman in relationship, as a vital technology for change. In her examination of Steppenwolf Theatre Company's 2000 revival of *One Flew Over the Cuckoo's Nest* and its critical response, Alexis Riley introduces the notion of "mad performativity," or the embodied movements, behaviors, and vocalizations acknowledged within a given historical moment as conveying madness. As Riley demonstrates, critics deployed clinical language in their *Cuckoo's Nest* reviews to authenticate or invalidate the actors' mad performances, suggesting that "mental health professionals, artists, and critics collaboratively author mad history through engagement with performance" (pg. 38). The chapter concludes in a call to approach theatre criticism as care, with Riley foregrounding accountability to mad communities and the expansion of cultural competencies as ways of advancing an anti-ableist critical practice. Choreographers and performers Kasi V. Aysola and Madhvi J. Venkatesh next detail the process by which they developed *Through Fish Eyes*, a 2019 dance production aimed at activating audiences' empathetic responses to human-caused marine degradation. In it, the dancers' embodiments, rooted in the Indian dance form of *Bharata Natyam*, variously represent both the wonders of marine life (whales, coral reefs, turtles) and the forces threatening the ocean's health and stability (scientists, fishermen, plastic pollution). Aysola and Venkatesh contemplate *Through Fish Eyes*' capacity to inspire audiences to safeguard the world's oceans. In "Laboring the Medical: Female Bodies for Sale on the Contemporary Stage," Gianna Bouchard examines three contemporary UK plays by women in which women's bodies labor in clinical contexts, as test subjects and international gestational surrogates, and thereby become biomedical commodities. Such productions, Bouchard argues, foreground the "ethical and social dilemmas facing women entangled within this global economy" (pg. 77) as well as biocapitalism's dependence on femininized (and racialized) clinical labor.

Themes of representation and identity are explicitly taken up in Part Two: Challenging Traditions through the Science Performance. The chapters in this part explore science performances enacted outside the laboratory by scientists whose communities are underrepresented in science fields as well as the potential impacts of prioritizing more inclusive scientific processes. In Chapter 4, "Spooky('s) Action at a Distance: Remixing Science and Performance in the Planetarium Show," Mike Vanden Heuvel proposes the remix as a method for teasing out the philosophical and aesthetic complexities that materialize when science and art converge. Using the example of Paul D. Miller's (aka DJ Spooky) remix installation *The Hidden Code*, Vanden Heuvel builds upon Michel Serres's theory of *parasite* as a means of destabilizing art-sci discourse. In her practice-based chapter, "Using Short Digital Films to Counter Stereotypes about Scientists of Color and from Marginalized Backgrounds," Mónica I. Felíu-Mójer describes a process for communicating marginalized scientists' stories that challenges the deficit perspective pervading media representation. This chapter follows the documentary filmmaking process at work among Felíu-Mójer and her colleagues at the Science Communication Lab's Wonder Collaborative, which produced the "Background to Breakthrough" series of short films. Felíu-Mójer's chapter expands upon the Wonder Collaborative's rationale, the scientists' stories, and recent research on diversity within STEM fields. In Chapter 6, Felipe Cervera asks you, the reader, to consider the ground beneath your feet the next time your gaze wanders toward the sky. Cervera queries the implications that colonial histories of astronomy, astronautics, and nomenclature hold for Indigenous imaginations and cultural traditions, using the performance of Zapotec *muxe* artist-anthropologist Lukas Avendaño as a model for this ground-up cosmology.

How have histories of conquest been enacted, at least in part, in the name of science? Part Three: Revising the Art-Science Repertoire contains chapters examining how science has informed acts of European conquest and the rise of national liberalism in European and Euro-American occupied lands. With the chapter "Performing and Negotiating Imperialism: Science, Agriculture, and Food in Puerto Rico," Teófilo Espada-Brignoni introduces the field of agriculture as central to the American colonial project. Mangoes and the island's Agricultural Experimental Stations are Espada-Brignoni's historiographical objects, through which he interrogates the scientific performances of the US acquisition of Puerto Rico following the Spanish-American War. Derek Gingrich's focus on Bertolt Brecht's various iterations of the play *The Life of Galileo* elucidates the philosophical and scientific underpinnings of the rise of German nationalism. Gingrich proposes that Brecht's 1938 *Galileo* was more than a "naïve" celebration of the possibilities of atomic science but a critique of anti-intellectual nationalist trends "endemic to German academia amid the country's interwar identity crisis" (pg. 180). Also finding twenty-first-century comparisons to performances of interwar politics, Emily Klein links the Federal Theatre Project's science-focused Living Newspapers and the recent efforts by Pittsburgh, Pennsylvania's Squonk theatre company to combat public distrust in science during the Trump presidency. Detailing the imbricated nature of science, performance, and politics through her paired case studies, Klein's chapter presents theatre's community-building and educational capacities as radical antidotes to the Trump era's anti-science rhetoric and political divisiveness. In the second artists' roundtable, playwright Lauren Gunderson, actor William DeMeritt, and dramaturg Martine Kei Green-Rogers reflect on the premiere production of Gunderson's *The Catastrophist*, which was written, rehearsed, performed, and filmed during the height of the Covid-19 pandemic. A digital production for the lockdown era that depicts a virologist specializing in pandemics, *The Catastrophist* met audiences where they lived, both literally and figuratively. Gunderson, DeMeritt, and Green-Rogers discuss how *The Catastrophist* came to be, the opportunities and drawbacks of remote theatre-making, and why science performances matter.

We are pleased to include five creative interludes across the entirety of this volume. Together, these monologues, performative poems, and play excerpts model how dramatic texts themselves work as a technology to challenge assumptions, spark imagination, and counter oppressive hegemonic narratives—about science! In a melding of dramaturgy and biology, Diane Stubbings's *Variation for Three Voices on a Letter to Nature* is a performance text that emerged from practice-based research into the application of biological processes and prototypes to the generation of dramatic texts, using the Watson and Crick letter to *Nature* (1953) and Samuel Beckett's *Krapp's Last Tape* (1958) as generative material. Claudia Barnett's "Please Let Me Shoot You: A Monologue" dramatizes an oceanic biologist's experience of climate change anxiety (CCA) and her individual contributions to understanding one aspect of our anthropocentric moment. Chantal Bilodeau's monologue "Mother" gives dramatic voice to our ailing planet, calling into question the visioning of human-planet relations through a parent-child rubric. Kim TallBear has offered three of her Critical Polyamorist 100s, a poetic discipline that she discusses in the Climate, Colonialism, and Performance Roundtable. These short poems directly challenge the cultural dominance of heteronormative human relations as part of colonial traditions of harm. A short excerpt from Lauren Gunderson's *The Catastrophist*, a solo show based on the life and work of her virologist husband Nathan Wolfe, weaves together threads that run throughout this volume: of personal and social identity, the creative process of theatre-making, and science issues that—though perhaps difficult to discern due to historical, experimental, or cultural circumstances—are of great consequence to those of us living on Earth in the twenty-first century.

Notes

1 William Shakespeare, *The Tragedy of King Lear*, Folger edn, ed. Barbara A. Mowat and Paul Werstine (1606; New York: Simon and Schuster, 2015), 124.

2 Other historical thunder runs remain in use at London's Her Majesty's and The Playhouse theatres. See "The Thunder Run," *Bristol Old Vic*, https://bristololdvic.org.uk/archive/thunder-run.

3 There is still some debate as to whether wooden balls, cannonballs, or round iron weights were used by Georgians; the 2016 production used wooden balls after testing different objects for sound quality.

4 The Georgian acoustical technology likely built upon earlier tempestuous sound effects involving "drums and rolling cannonballs to produce thunder," even on the Jacobean stage (Mary Thomas Crane, "Optics," *Early Modern Theatricality*, ed. Henry S. Turner (Oxford: Oxford University Press, 2013), 265). Keith Sturgess describes the sound effect at Blackfriars for the Jacobean production of Shakespeare's *The Tempest*: "Squibs from the upper level of the Blackfriars façade or a resin box provided the lightning and the thunder was mimicked by drums in the tiring-house or music room or by cannonballs rolled in a thunder run, perhaps a combination of the two." He cites Ben Jonson's disdain for this kind of aural theatrical effect as expressed in the 1616 prologue to *Everyman in His Humour*. See Keith Sturgess, *Jacobean Private Theatre* (London: Routledge & Kegan Paul, 1987), 81, 217n.10. However, Paul Mezner refers to *Lear*'s storm sound effects as having been produced by thunder sheets in the essay, "In the Event of Fire" (*Moving Shakespeare Indoors: Performance and Repertoire in the Jacobean Playhouse*, eds. Andrew Gurr and Farah Karim-Cooper (Cambridge: Cambridge University Press, 2014), 178).

5 Sophie Chiari, *Shakespeare's Representation of Weather, Climate and Environment* (Edinburgh: Edinburgh University Press, 2019), 157. *Lear* engages in the early modern debate over judicial astrology, asking to what extent do natural phenomena dictate human actions and whether human emotions can bring about severe weather or natural disasters. For more on the nuances of early modern astrology and astronomy, see Crane's "Optics"; Hugh G. Dick, ed., *Albumazar: A Comedy 1615, by Thomas Tomkis* (Berkeley, CA: University of California Press, 1944); and Vivian Appler, "Among Actions, Objects, and Ideas: The Telescope in Thomas Tomkis's *Albumazar*," *Comparative Drama* 50, no. 1 (2016): 81–105.

6 Gloucester declares Nature's wisdom to sometimes be obscured to human perception. He draws his fears from his reliance on astrological prognostication: "These late eclipses in the sun and moon portend no good to us. Though the wisdom of nature can reason it thus, and thus, yet nature finds itself scourged by the sequent effects" (1.2.101–6). The other side of this coin of observable/hidden natural causes is expressed by Edmund, Gloucester's bastard son, who declares, "we make guilty of our disasters the sun, the moon, and the stars, as if we were villains by necessity, fools by heavenly compulsion, knaves, thieves, and treachers by spherical predominance, drunkards,

liars, and adulterers by an enforced obedience of planetary influence, and all that we are evil in by a divine thrusting on" (1.2.120–6). Edmund knows his own capacity for villainy. With these contradictory statements, the playwright demonstrates the early modern stage's capacity to work as an inquisitive technology that might uncover not only the depths of human nature but natural mysteries as well.

7 Chiari, *Shakespeare's Representation*, 160 and 162. Chiari acknowledges that *Lear*'s premiere was probably not at the Globe but at court, at Whitehall. However, Yoshiko Kawachi's *Calendar of English Renaissance Drama, 1558-1642* places the earliest possible date of performance of *Lear* at the Globe in 1604 but no later than 1606 (Yoshiko Kawachi, *Calendar of English Renaissance Drama, 1558-1642* (New York: Garland Publishing, Inc., 1986)).

8 "Thunder Run Experience."

9 The field of mad studies is discussed in detail in Chapter 1, "Staging the Mad Past: Performance, Criticism, and Historiography in Steppenwolf Theatre Company's *One Flew Over the Cuckoo's Nest*," by Alexis Riley.

10 Andrew Sofer, *Dark Matter: Invisibility in Drama, Theater, and Performance* (Ann Arbor: University of Michigan Press, 2013), 7.

11 See Meredith Conti, *Playing Sick: Performances of Illness in the Age of Victorian Medicine* (London: Routledge, 2018), 188–95.

12 Sofer, *Dark Matter*, 3. Although we know its effects, the graviton (a particle that communicates the force of gravity between objects) was long one of the undiscovered particles theorized by quantum physics. Gravitational waves have only recently been observed by physicists. A good popular source describing the history of gravitational theory is Natalie Wolchover's "Gravitational Waves Discovered at Long Last," *Quanta Magazine*, 11 February 2016, quantamagazine.org.

PART ONE

Performing Human and More-Than-Human Relationships

Performance, Climate, and Colonization Roundtable

Claudia Barnett, Chantal Bilodeau, David Geary, Kirsten Lindquist, and Kim TallBear

In June 2022, a group of performance artists, playwrights, and scholars convened over Zoom to discuss their work as it pertains to issues of identity, colonization, and climate. The panelists identified thematic overlaps and connections in their artistic and scholarly pursuits, together highlighting the ways that theatre, performance, and intentional practices within daily life can help us to shift perspectives; reframe relationships between humans, nonhumans, and the natural world; and cultivate sustainable habits for life on Earth.

Vivian Appler: Thank you all for participating in this panel.

Claudia Barnett: I'm Claudia Barnett. I am a professor at Middle Tennessee State University. I live in Lascassas, Tennessee, where it is almost 100 degrees, and I am here with my dog Lucy. I write plays, some of which are about science and nature, and it's my pleasure to be here and to meet all of you.

Kim TallBear: I'm Kim TallBear, she/her. I am a citizen of the Sisseton-Wahpeton Oyate, which is a Dakota people in the eastern side of South Dakota. Our historic homelands are Minneapolis and Saint Paul . . . I am a professor at the University of Alberta in Edmonton, in Canada. I'm a Canada Research Chair in Indigenous Peoples, Technoscience, and Society. And, I run a summer Indigenous Peoples Genome training program that contextualizes genomics within decolonization and instructs aspiring scientists on how to do that. I'm also the co-producer with Kirsten Lindquist, here, of Tipi Confessions, which we'll be talking about today.

Kirsten Lindquist: Tânisi. Hello. I'm Kirsten Lindquist. I'm a citizen of the Métis Nation of Alberta . . . I use she/her/they/them pronouns. I'm a PhD student in Indigenous Studies at the Faculty of Native Studies at the University of Alberta, co-producer of Tipi Confessions, and I'm also a research assistant for Dr. Kim TallBear's research-creation lab: RELAB.[1]

Chantal Bilodeau: Hi, my name is Chantal Bilodeau. I am originally from Montreal but have been living for over twenty-five years in New York City, which is the land of the Lenape people. I'm currently in Anchorage, Alaska. I am a playwright, the artistic director of the Arts & Climate Initiative, and a co-producer of Climate Change Theatre Action.

VA: The first question is about origins. All of you write plays or create performances that engage questions about environment and intersectional oppressions. Has this perspective always been intentional or has it evolved out of your practice? How does your work actively intervene into these problems?

KT: Regarding Tipi Confessions, a sexy storytelling show that Kirsten and I founded in 2015 here in Edmonton, the show definitely started out looking at what does decolonizing sexuality look like?[2] Interrogating the role of sexual violence and the hypersexualization of Indigenous women in the colonial project . . . We were definitely focused on . . . I try not to use cliched words, but, celebrating sexuality, looking at sexuality as joyful, as perhaps healing. Our other co-producer, Tracy Bear, who's a professor at McMaster University, teaches an Indigenous erotica class, and she said when she googled Indigenous women and sexuality back before 2015, all she got was trauma and violence, and that's really heartbreaking . . . So, we really tried to address a lot of that and . . . facilitate an environment where the audience members and the performers can . . . look at the healing, and the good, and the joyful, and the empowering things in sexuality.

That said, because we are Indigenous scholars, and thinkers, and artists, and community (I mean, many people come to our shows who are part of the community who are not necessarily academics), I always say some of our best theorists are living in community, right? And, they're thinking sometimes from out of Indigenous languages. They're thinking from out of their decolonial practices in their lives. We come from these relational frameworks and relational ways of viewing the world.

So, our goal was always . . . to move . . . beyond human sexuality, to think about good relations beyond the sexual . . . So, when we're thinking about good relations, we can think very easily about our more-than-human relatives and our very recent show, which . . . Kirsten will get into and further answer, was an online show and that is where I saw us fully realize this kind of melding of good relational sexual relations among humans and then good relations with our nonhuman relatives and the land . . . That's exactly, I think, what we envisioned seven years ago and it's coming to fruition now (Figure 1).

CB: I want to say thank you for the question because, when I read the question, I thought, "I do that? I didn't know I do that." Then I started looking at my plays and I was like, "I do that!" And, it was very funny to me because I wrote criticism before I wrote plays and the idea that I have no idea what I'm doing was really amusing to me. So, thank you.

ChB: I trained as a playwright and for the first several years of my career I wrote plays about a variety of things. It was a trip to Alaska in 2007 that made me change direction. This was a year after Al Gore's documentary, *An Inconvenient Truth* (2006),[3] came out and climate change was a lot more present in the mainstream conversation . . . Of course, in Alaska, there was no debating whether it was happening because people were witnessing changes every day. Glaciers were retreating and deflating, migration patterns were changing, and species were moving north. After that trip I started thinking, "Maybe I can bring these concerns into my work." I originally thought I was going to write one play. And I did write

A Recipe for Seasoned Sex

With the right ingredients, sex, like love, can age well. With age one might learn to take risks, to not forego love's richness out of fear that its spice might burn. You are in luck! Here is a hard-earned sex-culinary secret recipe for cooking up the most succulent, smooth-on-the palate Seasoned Sex.

1/2 cup tolerance for different bodies fitting together differently sift together with ½ cup curiosity
fold in ½ cup courage
add 3 heaping tablespoons of safer-sex knowledge
cut with 2 teaspoons compersion

NOTE: bake at 37C /98.6F preferably for 2-3 hours (quickies can do the trick, but a slow bake is preferred) in a possessiveness- and shame-free kitchen.

If eventually your Seasoned Sex cools a bit too much, remember that like love, it might mean it has sufficiently nourished you. Lovers should not be afraid to test the nourishment of different kitchens. Let your tongues explore new tastes. Just like a meal come to an end is not failure, neither are relationships ending or transitioning failures. Sometimes our desire for certain spices changes over life's seasons.

**FROM THE KITCHEN OF KIM TALLBEAR
(AKA THE CRITICAL POLYAMORIST)**

FIGURE 1 *"A Recipe for Seasoned Sex." Words by Kim TallBear, image by Kirsten Lindquist.*

the one play, but once I got started I couldn't stop. Everything I was learning from scientists and coast guards and people who I normally wouldn't have access to was so fascinating that I decided to write a whole series of plays.

Meredith Conti: Our next question is about otherness, marginalization, and erasure. How have relationships with bodies and persons that have been othered—human, non-or-more-than-human, technological—influenced your playwriting and performance-making? Or, how has your creative practice brought an awareness to your own or your audiences' relationships with these kinds of categories: others, otherness, and the process of othering in the practice of daily living?

KL: I also work with Dr. Lana Whiskeyjack on a strengthening relations project . . . We're working with two-spirit LGBTQIA+ youth . . .[4] That has also shaped the way that I've been thinking about . . . where Tipi Confessions is going . . . Our most recent online production was with the Toronto Queer Film Festival, and it was online . . . Thinking about . . . how not only Indigenous women and girls but also our two-spirit relatives, how the violent imposition of colonial myths have shifted the way that . . . women, two-spirit, gender non-binary people have been connected to the land—their land-based practices—and how that's related to resource extraction and climate change. For me . . . unraveling those complex and entangled narratives . . . How do we reclaim or restore those narratives with youth? Not only the youth on the land, but then listening to the stories when we're not able to visit on the land . . . Because of the pandemic, we had to move to online spaces and . . . with this online production we saw this very diverse expression of Indigenous women and two-spirit, non-binary people really . . . wrapping this up in a great way to show that these stories can be told in the online space as well.

ChB: Before I started writing about the climate crisis, I had a strong desire . . . to write about nonhumans. I wasn't aware of it at first, but when I looked back at my plays I realized, "Oh. yeah, there's always a nonhuman element." Sometimes it's an animal. Sometimes it's someone who hasn't been born yet. Cats. Sometimes it's sea ice. When I started writing plays about the environment and the climate crisis, I became . . . more intentional about this as a way to put humans in the middle of a web of life as opposed to presenting them as the dominant species. In addition to decentering humans, including nonhuman characters gives me the opportunity to present a different perspective on humans. It's still me writing, but it opens up the possibility of having characters reflect on what we humans do.

CB: I was interested, thinking about this question, about the otherness of women in general—the irony that half of humans might be othered by gender. Somehow, I have equated otherness with nature in my plays. And that, again, seems like an irony, given that here we are among nature, and yet so many people seem so distant from nature. Women having a bond of otherness with nature is an issue that I want to contemplate further.

VA: This next question is about visibility and invisibility. All of you use science-oriented performance or playwriting to make visible the invisible or hidden aspects of the human experience: hidden natural histories, untold stories of underrepresented scientists, invisible pollutants in our environment, and the invisibilized harm of colonization, hiding often in plain sight. Please share more about a specific example of a play or performance that speaks to the power of theatre or performance to raise awareness about these environmental issues that are deliberately or inherently obscured in the course of the everyday.

CB: I have a character who's a waterfall in *Kingdom (a play about Snow White and climate change)*—wrapped in Christmas lights and walking on stilts.[5] When other characters see her (they often don't), they see a way to make money, or they see a way to power their factories, or they just see something for their own use . . . I see her as female. She is the Rjukansfossen, a waterfall in Norway, and she's tall and glorious and a sight for hikers to come see in the nineteenth century. Of course, she's polluted by the end of the play. But, at the start of the twentieth century someone buys her—and I was struck by the idea that you could buy a waterfall—and uses her to power a fertilizer plant so that crops could be made larger. So, it was kind of changing one form of nature in order to change another form of nature. And then there was a byproduct of this whole process, heavy water, which is an isotope (D_2O instead of H_2O) that was later used to make nuclear weapons—again, this idea of taking something natural, changing it slightly, and potentially causing a great deal of harm as a result. Or, on the flip side: The valley of the waterfall gets no sunlight in winter, so an artist installed mirrors on the town square to shine light. My play, which has "climate change" in the title, isn't about a conventional kind of climate change, but instead about this one spot in the universe where things would get changed again and again and again. And the waterfall's there the whole time as a constant force, except that *she's* being forced constantly to become other things.

KT: This isn't specifically about environmental issues that are deliberately or inherently obscured, but it is about other kinds of obfuscations . . . We did a show at the University of Washington . . . that brought Tipi Confessions together with our Summer internship for INdigenous peoples and Genomics (SING) . . . We proposed to these scientists who organized the summer program in the United States that we would like to do a sexy science confession show and they were, it turns out (we found out when we got to Seattle after planning it the whole year), nervous about it. They didn't know what freaky thing we were going to do.

Samantha Archer, a grad student at the University of Connecticut who uses ancient DNA to study long-ago societies and inheritance of genetic changes due to social conditions, did a story called—it's so hilarious—called "Authorized Personnel Only." Sam did their undergrad in gender studies at the University of Texas . . . and then was also doing ancient DNA work. Anyway, so [Sam] talks about growing up bisexual but . . . didn't know they were bisexual—this is all very funny—until scientists proved bisexuality existed. Before that they were just promiscuous. And they talk about, "I wasn't good at math and I was gay, so of course I couldn't be a scientist. In Texas, scientists . . . aren't sure gay people should be scientists. And the rest of the country doesn't think Texans should be scientists." So, because Sam was gay and a Texan, there's no way Sam could do science (and bad at math).

So . . . Sam goes through the whole piece and just makes fun of the DNA lab. Totally makes fun of it. Demystifies it. Then, Sam started talking about the, what's the machine that spins DNA I forget.

VA: A centrifuge?

KT: Yeah, centrifuge! Like, that's one of the best sex toys ever invented—the thing vibrates! So, Sam's talking about all the equipment that could be used as sex toys, and she's like, "These are just really expensive sex toys." And, then she talks about getting off on it, like, everybody gets off on science. I think it was hilarious and totally demystified it. But, Sam's also an accomplished scientist now, so . . . really, I love that.

KL: We partnered with QTPOC [Queer and Trans People of Color] in Winnipeg Treaty One in 2018, which was online. It was a Halloween-themed show, "Sex at The End of The World," and it was intended to challenge settler narratives of Apocalypse. So, we had the performers really challenge that this isn't the first apocalypse. Some of them were immigrants to Canada and some were Indigenous, but it was this idea that . . . we've been surviving apocalypses for centuries, so when we think about climate change . . . how are Indigenous people, Black people, people of color . . . can their restoring—especially through ideas of gender and sexuality—provide some sort of resistance to the way that bodies connect to these destructions in land? How can telling these stories create a possibility of re-kinning with the land despite the violence of disconnection that we've experienced for centuries?

So, it's not just addressing that we're also in another apocalypse, it's creating alternative futurities . . . I think that futurity also runs through many other shows that we hosted. One in November 2019 was part of our RELAB Research Symposium. So, this is another way of thinking about unsettling . . . the future of science . . . what is it in our own stories, our creation stories, our transformation stories that gives space to think differently? What is futurity for addressing climate change, addressing pollution, addressing . . . structural issues that specifically two-spirit, LGBTQIA+ people need in terms of homelessness and incarceration that is also connected to climate change? It's always very much entangled . . . Each time we have a production, that gives another iteration of our futurity in response to what is going on right now.

One specific performer that we had for our Vancouver show, Mother Girth, their debut performance was "Water Spirit's Retribution."

KT: What struck me about Mother Girth was . . . they came out on stage in this gorgeous . . . blue sequiny, kind of like water . . . powwow music . . . They came out doing what we would normally call men's fancy dance, which is the end of the night. If you go to a powwow, it's super gender binary . . . and age stratified. You start with the little kids, boys and girls. Then, you go teenagers, boys and girls. Then, you do men and then you do seniors.

So, Mother Girth comes out and is in a dress sequined like men's fancy dancing, which is the flashiest, most wild, end of the night, scream-raising performance when you go to a powwow. It's, like, at 1:00 AM . . . it's so wild and fast . . . I started crying, I was, like, "Oh my God!" It was just . . . Really, like, these macho guys do this. I loved it.

I liked it for that, for the challenging of the gender binary . . . You get down on the floor. You spread your legs. It takes a lot of agility and just a complete comfort with your body and opening your body up. And that's why I liked it . . . I was so blown away, yeah. And "Mother Girth," too . . . I thought that was such a great play on words.

KL: Yeah . . . Mother Girth was representing the water spirit, the anger, the rage of the destruction of these waterways through resource extraction. To connect that with the audience. To show the feeling of rage of the water spirits, I think, in addition to challenging, or queering what we think of traditional dance, I think being able to be a conduit—that many of our performers are a conduit—to bring expression, to bring the emotion of maybe what people are not listening to: our more-than-human relations.

ChB: I can talk about my most recent play, No More Harveys . . .[6] In 2017, Hurricane Harvey hit Texas and Louisiana and a couple of months later, Harvey Weinstein was accused of sexual abuses. For me, it was a big like [*she makes an "Aha!"* gesture], and I thought, "Isn't anybody else seeing the connection?" I wrote a play where the world is full of Harveys, which in the play are defined as . . . "men who are fed and fattened by money and power, and prey on vulnerable people. Especially women." The play tries to . . . show the root cause of issues that we often consider separately, like environmental destruction, gender-based violence, and health issues.

In the story, a woman goes on a journey across the country . . . She's running away from her own Harvey, her husband, and tries to make sense of all of the other Harveys in the world. In order to do that . . . she seeks guidance from what she calls "the whale from fifty million years ago." Whales used to be land animals and fifty million years ago they lost their teeth, lost their legs, and moved into the ocean. For the woman, that's a feat of adaptation and survival.

In recent years, the climate movement has focused on making connections between what were once seen as separate issues: racism, economic disparity, climate justice. I wanted to write a play that would bring these things together, create a vivid image so everyone could understand how they are connected.

MC: We started to think about how your individual commitments to ecological preservation and against climate breakdown might show up in your creative works. Do certain dramatic structures, theatrical styles, or ways of using bodies and voices better serve your climate-minded purposes? Are you drawn to particular visual or aural landscapes? Does this impact with whom and how you choose to collaborate, or, how you travel and communicate for this work? Do you write in a way that encourages, perhaps, a low carbon or green theatre practice? This might also tie into decolonization of theatre praxis.

KT: I'll just say the sort of the obvious thing, but I know this is complicated . . . I needed a break from air travel. Really . . . I was one of the worst committers of . . . superfluous air travel . . . I was doing well over 100,000 miles a year. It was hard on my body, and it's terrible for the planet, and a waste of time. So, I've really embraced the Zoom era.

That said . . . I don't know what the politics are around the servers used . . . to support all of this . . . The online show we did, I think, was a real breakthrough. We could not have done it without Toronto Queer Film Fest. The amount of time, and labor, and money that went into that and their technical capacity. So, that's the thing, right? We don't all have those technical capacities. We don't all have the institutional resources to embrace this era. And, so, that's not really, probably, ultimately the solution to working more locally.

I also moved to Edmonton because I wanted to be on the prairie, in a city where I see Indigenous people as everyday people. I got tired of living in places like San Francisco, or Boston, or Austin, Texas, where we're nothing but fetishized. Here, I can see a Native guy who's a mailperson. There's a young woman at the wine shop. My next-door neighbor . . . That's where I want to live. Where we're everyday people. So, that also enables . . . the kind of work you can do, locally, on Indigenous issues in Edmonton is phenomenal. It is so much better than anything I could have done in San Francisco, Austin, or Boston, which are probably widely held as being more cosmopolitan cities. But, there are not nearly the percentage of Indigenous people there. There's not the thriving Indigenous institutions and cultures. . . . For me, it was a decision to come to a place where I could do this work without having to also travel all over the place, which is unsustainable for my body as well as for the planet.

I hadn't thought of that move in terms of an environmental move, but, I think it really does lead to that as well . . . It's not only the broader ecology, but it was the ecology of my own body and life that was unsustainable. I was breaking down. I was getting unhealthy. I was in pain . . . I am much healthier now that I'm much more local. But, I had to come to a place that had what I needed . . . We may all have to make those decisions at some point. We can't be everywhere. We can't be the cosmopolitan subjects that we thought we could be. I just don't think it's sustainable. And, online will only take us so far.

KL: I can add to that in terms of thinking about how Tipi Confessions are about both body sovereignty and larger Indigenous sovereignties . . . part of a network. When I imagine this in terms of the local network, but broader, when we do international shows . . . this is a network of elders, and researchers, and practitioners, knowledge keepers, educators, doulas, sex workers, storytellers, artists, performers, tech people, and scientists . . . Knowing that there is this information and knowledge generation exchange where people can show up within this network . . . Moving forward, it's strengthening that network and not reinventing anything new, but, understanding that people have developed a lot of movements and spaces where they are addressing, for example, our Indigenous women and two-spirit, LGBTQIA+ relatives on the frontlines, defending lands and waters.

And, what is the aftercare? As much as we tell those stories . . . what are we doing in the aftercare and providing support for these folks in the local environment? We see these . . .heightened responses, like Wet'suwet'en land and water protectors and the tiny house warriors in the mainstream news,[7] but, what is also happening at the local levels as well?

ChB: I'm a co-organizer of a project called Climate Change Theatre Action. Every other year, we commission fifty playwrights from around the world to write a short play about an aspect of the climate crisis based on a theme. Then, we take this collection of plays and we make them available to anyone who wants to organize an event in their community within a certain time window. Although this model existed . . . prior to us doing this, we've adapted it and grown it . . . The traditional production model for theatre is that you create a show and then you tour it, which is very expensive and carbon intensive. The Climate Change Theatre Action model allows for cross-pollination between nations; playwrights and participants are from all over, but there's no travel involved. The plays circulate by email and the events are organized locally. The project offers a different way of being local and global at the same time while minimizing carbon emissions.

VA: This last question has to do with action, performance, and performativity. Performance is action—an active process capable of responding to all kinds of social and environmental issues. Likewise, the Earth is always moving, changing, responding to internal and external impacts and stimuli. But the Earth's motions are often occluded by human cultural performance. In your experience, how have acts of performance proved to be capable of resisting colonial legacies and environmental harm, and to what extent do you approach performance creation as a means of moving with the Earth and thereby altering or enhancing our human relationship with it?

KT: I write these Critical Polyamorist 100s.[8] The 100s are great writing practice . . . I write these exactly one-hundred-word vignettes . . . The discipline of that has really been good for me. I love being disciplined by one hundred words. It's very generative for me and very . . .exhilarating. It feels good. There's probably some kink metaphor in there . . . I just love the discipline . . . Writing those, like our shows, very quickly for me turned from writing about polyamorous human lovers to talking about my relationships with place . . . I started writing those when I was at the University of Texas a year or two before I moved up to Edmonton, in 2015.

They're always simultaneously about a human lover and about place. I'm a river person and I'm a prairie person. And I was really able to explore that. I think writing the 100s and focusing on my relationships with place helped me come to realize that, in fact, I am not commitment phobic, that I am very committed to the prairies and I'm very committed to this landscape. I have, in fact, chosen the prairies over a marriage. I've chosen the prairies over human loves. I will leave human loves, but I have to be in this landscape.

I came to understand that quicker by writing those 100s and performing them. So, that was really better than therapy and a lot cheaper, and a lot more exhilarating.

KL: Yeah, it's similar to Kim. I'm connected to the prairies. After moving to Victoria for one year, in 2015, I moved back. And that was the only time I realized . . . "Oh, I'm not an ocean person. I'm definitely a prairie person." So, when I came back, I committed to observing the seasons more, in a sensory, bodily connected way. I learned from Elmer Ghostkeeper, who has a book on the moons,[9] the Cree moons . . . Lana Whiskeyjack and Dwayne Donald have this practice of identifying your bodily connection with the moons.[10] Since 2015, I've been invested in a practice of keeping a journal about the changing landscapes and waters, the positions of the constellations. Over a long period of time, we can see why the moons are named the way they are, and so we just came out of Egg Hatching Moon and I was at home and I could hear the little magpies. The babies were hatching.

This is not just . . . knowledge that came before, but this sensory connection can also help us understand the intimate local impacts of climate change and how climate is shifting. This happens over a long period of time, and much of this information that we have here has been . . . observed over hundreds and thousands of years. So, this is just a very tiny kind of movement within myself to connect to the landscape . . . I did one performance—it wasn't specifically related to Tipi Confessions, but Tipi Confessions is a space that is an incubator for the way that I also do other creative performances—so I had this unitard that I screen-printed the Indian Act on, as a first iteration.[11] The last iteration of the performance, I developed moss pasties—I do dabble in burlesque from time to time—but, I had a performance where, as part of an Indigenous feminisms workshop, I had the audience come up and cut a piece of the Indian Act off my body, put it in the garbage can. Then I had a pile of dirt and bean seeds. And, so, we planted this together until I was naked except for moss patches and pasties. And . . . everybody had this bean seed . . . Sometimes, we don't get to know the impact of our shows or our performances, but I started receiving photos of the bean seeds growing. And, [I thought], "Oh, they're actually taking care of this." We think of the metaphor of the seed, of how this continues on in different cycles . . . One of the audience members grew beans from the bean seed, harvested, dried the beans, and then planted it for the next season.

CB: I thought this and the previous question were kind of daunting in that I just write; I don't perform. I write and then I hope somebody will someday perform. Being a playwright is difficult that way, but I do think, when writing, of possibilities. There is such a magic to theatre. And there is such a freedom to writing for theatre, and writing for creative people that you trust, and knowing that even if you write something completely insane, someone will figure out how to stage it. I also think there's something insane about living in the woods in Tennessee, and writing plays, hoping that they'll get put on far, far away. But I find it important to have that solitary ability to think about something like a waterfall in Norway that I've never seen, and then write about it here, and hope that it will be realized someplace else. Theatre has all these possibilities of collaboration, even among people who aren't together.

ChB: I'm not a performer either. But, of course, I do think about performance when I write. Last year, we put on a show that was performed outside. It was several short pieces by different playwrights and I wrote the connective tissue that held them together. It says something different, I think, to perform outside, in the environment, when you're talking about the environment. It's very different from performing in a black box where all of the production elements are controlled.

    I, like any writer, write to make sense of the world, hoping that my writing will . . . resonate with other people. The writing process is a way to train myself to be better at speaking for all the things I'm witnessing, all the things I'm feeling. I have given myself the opportunity to connect with the environment and with how people are affected by the environment more strongly than if I was just going about my day-to-day life doing something else. I think I have created something that I needed for myself, which is to move with the Earth in much closer connection than I would have otherwise.

VA: In closing . . . What is growing in your gardens? That is also a performance of everyday life that so many of us have a connection to.

KT: Well, now that you mention it, I only have decorative plants. I don't have anything I can eat other than edible flowers for cocktails. That's the only thing I grow that I can eat 'cause you can't get them in Edmonton—it's too country to have cocktail flowers at the grocery store.

KL: I just have house plants in my apartment in Edmonton, but I did go home and my parents have a full garden: a lot of carrots, beets, potatoes, herbs, flowers . . . When I go walk in the river valley you can see that the Saskatoons are coming up, too. I try to . . . find the places before the birds and the other critters will get them.

Just as we were wrapping up, playwright David Geary joined the Zoom. Kim and Kirsten had to leave due to prior engagements, but we continued the conversation with David for a bit.

David Geary: Kia ora. How's it going? Sorry, I'm late. My pronouns are he and they and I'm on the Tsleil-Waututh, Musqueam, and Squamish territory . . . I prefer to say colonial and occupier now as opposed to settler 'cause I think settler sounds a bit too mild. I come originally from Aotearoa, New Zealand. I am of the Taranaki Nation or iwi. . . . I don't write many plays and I don't actually spend that much time in the theatre . . . My life is dominated by film and television.

VA: We had just asked what's growing in your garden.

DG: Rocket a.k.a. arugula, potatoes, spinach, parsley, and slugs.

CB: I have a zillion daylilies currently in bloom. They look awesome and the only thing that I have that's edible is herbs. It's a little bit late in the season for edible flowers, so I'm jealous of Kim, because it's too hot for pansies in 100 degrees, but we do have blueberries that are never going to turn blue because the cardinals are going to eat them when they're still pink.

VA: We had been talking about movement, the Earth, and artistic process . . . Another theme that has come up . . . is this idea of personhood. Who is empowered to speak and in what ways? How can playwriting make those voices heard? I'm also thinking about movements to bring personhood to natural bodies, specifically rivers.

DG: The Whanganui River in New Zealand did get personhood, although the river would probably say, "I don't really want to be a person. Look at people."[12] I would just say often with playwriting, you're wanting to represent voices that aren't in the mainstream . . . No one was speaking for the Tuatara, who's getting affected by climate change 'cause the heat of the ground dictates the gender of the thing.[13] So, when I found out that scientific fact, I was like, "I wanna somehow, in a trickster way, represent that how what we're doing is impacting these poor creatures that survived a lot of different colonial attacks."

CB: Well . . . I've already granted personhood to a waterfall . . . I agree that she wouldn't want to be a person. But she's cast as one, although she's tall.

DG: I would just say for Māori, there's a lot of different things that are treated as a person. Like, a mountain is a person, but also a canoe can be a person . . . Also, the houses . . . are named like they're people, and then they're designed like people. So, the roof is like the back and then the spine of the building, so that idea of them being animate is very strong.

When I think of Tanya Tagaq's work, the Inuit singer . . . her first album is called *Animism* (2014).[14] Everything is alive . . .

Often the Western world is just as weird, though. It's like they see it as objects to be exploited and things to extract. So, anything that gives other things real life, we would consider them animate and our brothers and sisters, as you, I'm sure you're aware, like the Buffalo being a brother who sacrificed their body for us so that we could eat.[15]

ChB: There's something about giving agency to nonhumans, making things animate. We're used to watching animals talk in Disney or animation films, where they're just stand-ins for people. But I think when we give agency to animals, or natural features, or houses, it helps us think about them differently.

CB: Maybe we aren't even conscious of how we think about them. I spent the weekend in the Bronx, visiting my parents. I was working in their garden. I have a garden, but I also have six acres that are surrounded by woods. I was in their garden, where they live very, very close to other people, and I suddenly realized I was talking to all the plants, and that people were watching me. I always do this, but I never thought about it because no one's ever watching me. I had not realized that's my relationship with plants. And suddenly . . . we become more aware of ourselves when we realize that we're being watched, which is kind of what theatre is. I was thinking about that as I was trying to be quieter this weekend.

DG: I'll give you an example, my friend Carman, who's of the Stó:lō Nation.[16] We were talking about hummingbirds. I would go, "Oh, it's really cold. Hummingbirds must be having a bad time." And he would come in and go, "Well, I'm looking after five hummingbirds, David. 'Cause whenever it freezes all the nectar freezes and they can die. So, I'm actually warming up nectar for them. I'm not just worried about them. I'm actually doing something." And I suddenly went, "Yeah, that's me . . . Oh, the poor hummingbirds." But . . . he has such a strong relationship that he is caring for them in a much more active way 'cause he realizes hummingbirds are really important. And . . . symbolic. One of my students told me if you hear a hummingbird but don't see it, change is coming but you don't know what it is. If you see a hummingbird and you hear it, change is coming and you know what it is.

That is a very different relationship to, "Oh, look at the beautiful hummingbird," just seeing it as something of beauty. He had a whole story and a thing about what engaging with the hummingbird was. In the same way that . . . you talk to plants, I know lots of

Indigenous people talk to the crows . . . Carman would also go, "You know, they're talking to you, David, when you see the crows, so talk back." Like, they're going, "Hey, what are you doing? Where are you going? What's going on? What are you holding in your hand? Gee, your eyes look nice and juicy. If you fall over and die right there I'd probably eat those." When he told me that, I go, "Yeah, that's right. That's . . . the crow."

We don't see them as pests. They are seen as pests here 'cause they dig out people's gardens, the terribly colonial grass . . . as in, the American lawn, right? It's colonized everything. I really appreciate people who start those conversations with birds and animals, 'cause that shows respect . . . Then it doesn't matter what people think about you . . . I mean, look at the way that people talk to their dogs. Look at the way they talk to their pet cats, and then they kiss it and then it licks their face . . . Coming from a country boy whose dogs worked, I really find it kind of weird. But . . . the internet is full of people with cat videos and dog videos, so they have completely normalized having relationships with animals. They've just chosen a certain section of animals . . . and the others they can send to the slaughterhouse and eat as a nice bit of, as Morrissey would say, you know, and this meat you so "fancifully fry" caused the animal to be tortured and died.[17]

CB: I do talk to crows, but it's never respectful.

DG: It isn't? They have memories. They'll remember the person who spoke harshly to them.

I went to your Grand Canyon Park and I met a guy who was a crow expert, who is one of the few rangers who's going to be allowed to be buried in the park. He's actually chosen his gravestone already; its color is Crow Black. He has spent a long time collecting stories . . . I'm constantly amazed how smart they are, and how they, unlike a lot of other things, they don't spend their whole time just looking for food. They've got enough food, so they can play . . . They can be playful. He was telling me about the female opening up the rubbish bin so the male could jump in. Then the male would come and, "tap, tap, tap, tap, tap," when it was finished. And the female would come and open the rubbish bin for him to hop out. He would fly out with all the stuff . . . The Crow ranger just had all these stories about all the years . . . about a crow that helped a Labrador dog back to its owners. Just . . . Crazy, beautiful stories.

Have you ever seen them attack an eagle? The eagles are kind of like big B52 bombers around here. They find it very hard to maneuver. But they are, like, scavengers, right? They can be scavengers.

Here, this isn't a very good video, but it's one I took at a park near us. Do you see that? That is . . . an eagle just sitting in the tree. There's a crow right above it . . . So, the crows, when they worry that the eagles are going to come and steal their eggs and eat their chicks. So there goes the eagle being chased by two crows now, like, "Get out of here, you bastard!" It goes way out in the inlet. He's going to attack it. Bam! Gonna take it again. Bam! Then, there's another eagle turns up. And now there's gonna be another third eagle, right? So . . . I love seeing them attack them. They just go for it.

MC: I just saw something very similar to that in my neighborhood . . . My kids watched it for ten minutes, thinking what an interesting kind of battle going on above our heads, that if we hadn't looked up, we would have never noticed it.

DG: That's again what Carman would say . . . you're in this world, but you're not seeing it.

Notes

1 RELAB is a "place-based research-creation space," at the University of Alberta Faculty of Native Studies and organized by Kim TallBear. See *RELAB: Restory/Research/Reclaim*, re-lab.ca.

2 For the Tipi Confessions show schedule and archives, see *Tipi Confessions*, tipiconfessions.com.

3 *An Inconvenient Truth: A Global Warning*, written by Al Gore and directed by David Guggenheim, Lawrence Bender Productions and Participant Productions, 2006.

4 Lana Whiskeyjack is an "art actionist" of the Saddle Creek Cree Nation, Treaty Six Territory, Alberta. More on her multidisciplinary work can be found at *Lana Whiskeyjack: Indigenous Art, Art Actionist and Educator*, lanawhiskeyjack.ca.

5 Find more about *Kingdom* and Barnett's other plays at *Claudia Barnett, Playwright*, claudia-barnett.com.

6 *No More Harveys* is the third in Bilodeau's Arctic Cycle. Find links to these plays and Bilodeau's other work at *Chantal Bilodeau*, cbilodeau.com.

7 The Wet'suwet'en land and water protectors are dedicated to defending traditional lands from the extractive practices proposed and enacted within their territories. For more on the Wet'suwet'en people and land and water protection, see *UNIST'OT'EN: Heal the People, Heal the Land*, unistoten.camp; CBC News, "Wet'suwet'en chiefs arrive in Six Nations for 'landmark discussions,' starting eighteen-day tour," *The Canadian Broadcasting Corporation,* 3 August 2022; and Michael Luoma, "Collective Self-Determination, Territory and the Wet'suwet'en: What Justifies the Political Authority of Historic Indigenous Governments over Land and People?" *Canadian Journal of Political Science* 55 (2022): 19–39.

8 See Kim TallBear, *The Critical Polyamorist*, criticalpolyamorist.com; and Kim TallBear, "Unsettle," kimtallbear.substack.com. See also, Kim TallBear, "Prairie Relations 100s," *The Dalhousie Review* 100, no. 3 (Autumn, 2020): 356–7; and Kim TallBear, "Critical Poly 100s," in *Shapes of Native Nonfiction: Collected Essays by Contemporary Writers*, ed. Elissa Washuta and Theresa Warburton (Seattle: University of Washington Press, 2019), 154–66.

9 Elmer Ghostkeeper, *Spirit Gifting: The Concept of Spiritual Exchange* (Nanaimo Strong Nations Publishing, 2007).

10 Dwayne Donald, "We Need a New Story: Walking and the wâhkôhtowin Imagination," *Journal of the Canadian Association for Curriculum Studies* 18, no. 2 (2021), http://doi.org/10.25071/1916-4467.40492.

11 The Indian Act is a piece of Canadian legislature, established in 1876 and revised repeatedly since then, that "regulate[s] the federal government's relationship with the Indigenous people it recognizes" (Julie Burelle, *Encounters on Contested Lands: Indigenous Performances of Sovereignty and Nationhood in Québec* (Evanston Northwestern University Press, 2019), 15).

12 In 2017, the New Zealand parliament passed the Te Awa Tupua Act, "granting the Whanganui River system legal personhood," an attempt to curb the damage incurred by extractive practices and embrace a "legal framework rooted in the Māori worldview" (Jeremy Lurgio in Whanganui, "Saving the Whanganui: Can Personhood Rescue a River?" *The Guardian*, www.theguardian.com, 29 November 2019).

13 "Mōrehu & Titi" is a one-act play that dramatizes a conversation between a tuatara reptile and a shearwater bird. See David Geary, "Mōrehu & Tītī," in *Identity, Culture, and the Science Performance, Vol 1: From the Lab to the Streets*, ed. Vivian Appler and Meredith Conti (London: Methuen Drama, 2022), 159–69.

14 Tanya Tagaq is a singer/composer from Ikaluktutiak. *Animism* (2014) is available on her website, *Tanya Tagaq*, tanyatagaqu.com.

15 Brother Buffalo is an important figure in Indigenous American culture; he sacrificed his body so that humans could eat and have warm clothes in the winter. The buffalo slaughter of the nineteenth century was a part of the ongoing genocide of Indigenous North Americans. For an arts-research perspective of this issue, see Brenda Vellino, "Intervening in Settler Colonial Genocide: Restoring Métis Buffalo Kinship Memory in Amanda Strong's *Four Faces of the Moon*," *Studies in American Indian Literatures* 32, no. 3–4: 149–75. See also, Tasha Hubbard, "Buffalo Genocide in Nineteenth-Century North America: 'Kill, Skin, and Sell,'" *Colonial Genocide in Indigenous North America*, ed. Andrew Woolford, Jeff Benvenuto, and Alexander Laban Hinton (Durham Duke University Press, 2014), 292–305.

16 Carman McKay's art explores and encourages connections among the human and more-than-human world. Read about his community-based work in Peak Web, "Carman McKay on His Work, Inspirations, and Experiences," *The Peak*, 20 July 2018, the-peak.ca/2018/07/carman-mckay-in-his-work-inspirations-and-experiences; and Matthew Claxton, "Garden brings native plants to Langley school," *Langley Advance Times,* March 13, 2017, langleyadvancetimes.com/community/garden-brings-native-plants-to-langley-school/.

17 "And the flesh you so fancifully fry/ Is not succulent, tasty or kind/ It's death for no reason/ And death for no reason is MURDER," The Smiths, "Meat Is Murder," in *Meat Is Murder* (Sire Records Company, 1985).

Creative Interlude*

Critical Polyamorist 100s

Kim TallBear

Mistress of the Shears (08.29.19)

Promiscuous traveler, my hair's been run through by fingers from Cambridge to Denver, Berkeley to Tokyo, Auckland, Jakarta. Then I found her, my colorist domme, in a #YEG salon. Like Morticia with a burgundy gloss on her own dark locks. Black clad with powerful boots and erotic accounts, lips of steel purse: "I know you're growing it. But I'll take an inch." From behind, her fingers encircle my neck. I shiver. Our eyes meet in the mirror. When we rinse, her palms cradle and caress my head, her own creation of flesh and dye. Her hands drip blood wine rivulets.

Winona Quarantine (04.14.20)

I hunt cornstarch in cupboards, but yeast is my take. What isolationist bakers covet. In a stealth drive across the sunrise city, I take this offering to a love's snow-dusted porch, her prize to find in morning light. I binge on decades-old mournful music and sail back across the bridge, the ice river. Are these notes nostalgia for the poverty of 20? When there was less, but more to die for? The medicine man who named our girl, Sweet Song Woman, said a

---

*These poems and others from TallBear's Critical Polyamorist 100s series may be found at "Unsettle," kimtallbear.substack.com.

Winona is protected until her own first is born. I remain protected. I worry for my daughter.

River Quantum: Instructions for Babygirl (05.13.20)

¼ my ashes to be spread across the North Saskatchewan. When ice flows and is lit by sun at a bend in central Edmonton. ¼ to be blown across the heavy Mississippi. Back to ancestors at Minneapolis, St. Paul. 3/8ths shall be offered up to the Big Sioux. Next to the old powwow grounds, make sure! That river is akin to my grandmother who fished its waters on close-to-home days when she did not venture to the Missouri. 1/8th, the remainders of me, shall be emigrated to the Corrib, my tumultuous river love near Galway's cathedral where I cursed Columbus.

1

Staging the Mad Past

Performance, Criticism, and Historiography in Steppenwolf Theatre Company's *One Flew Over the Cuckoo's Nest*

Alexis Riley

Steppenwolf Theatre Company's 2000 revival of *One Flew Over the Cuckoo's Nest* was, by all accounts, spectacular. Throughout the entirety of the roughly two-and-a-half-hour performance, the twenty-two-person ensemble transformed their collective portrayal

---

I would like to thank Leean Torske at Steppenwolf Theatre, Morag Walsh at the Chicago Public Library, and Naila Holmes at the New York Public Library for vital research assistance. For reading drafts and sharing feedback, my gratitude to the editors as well as Lindsey Barr, Slade Billew, Alison Kafer, Stephanie Lim, Priya Raman, Molly Roy, and kt shorb.

of psychiatric inmates into an athletic event. Eyes blinked rapidly, faces squeezed into tight contractions, torsos rocked back and forth, shoulders curled inward into slight shivers, gazes darted to and fro, and mouths nibbled at fingers, presenting audiences with a kinesthetic collage of mental maladies—and a profitable one at that. Tickets for *Cuckoo's Nest* consistently sold out despite several run extensions; tours to London and New York soon followed, as did a Tony Award for Best Revival of a Play.

Despite these accolades, the production also garnered significant critique. In a *Chicago Tribune* article titled "Psychological Drama: 'Cuckoo's Nest' Is Dated—but Is It Dangerous?" critic Julia Keller argued that productions of *Cuckoo's Nest* problematically frame mental illness as "a self-induced response to an unpalatable world," a rendering woefully out of sync with contemporary psychiatry. "We no longer believe that mental illness is a device strategically employed by creative geniuses or a political gesture designed to subvert the dominant paradigm," the critic argued. "We know that psychiatric problems are caused by brain chemistry." She then contextualized this development within a larger arc of scientific progress, writing, "In the 17th and 18th Centuries, mental illness often was regarded as a curse from God; in the 19th Century, a sign of artistic genius; in the 20th Century, the result of oppression; in the 21st Century, the consequences of chemical imbalances in the brain." Concluding that "Art heals, but sometimes it heals the very wound it created," Keller ultimately located the source of her discomfort in the production's inability to distinguish between psychiatry's past and psychiatry's present, implicating the revival in the spread of potentially harmful medical disinformation.[1] In sum, *Cuckoo's Nest* was not simply bad theatre; *Cuckoo's Nest* was bad science.

Keller's review appeared in the paper's arts and entertainment section. It was not intended as science journalism, nor was it offered as a work of medical history. Yet, her article contains elements of both. Drawing on psychiatric theory, the critic interpreted the Steppenwolf production diachronically, transforming her performance review into a brief history of mental illness and its treatment. She did not limit her scope to the clinic; rather, she interpellated the theatre into her narrative, endowing performance with a therapeutic capacity to both "wound" and "heal." While notable for its interdisciplinary rigor, Keller's approach is hardly unique; as I discuss later in this chapter, other critics similarly relied on psychiatric theory and medical history in order to evaluate the Steppenwolf production. Mental health professionals likewise shaped this critical reception, weighing in with expert testimony and, in one case, penning a review. Together, these interdisciplinary responses point to an imbrication between clinic and stage, one in which performance both constructs the past and alters its meaning in the present.

This imbrication is the subject of this chapter. While the clinic and the stage are often considered separate domains, I argue that both mental health professionals and performance critics share an interest in embodiment, evaluating external signs as indicators of internal states. The Steppenwolf revival and the criticism it inspired, therefore, not only demonstrate this shared interest but also prompt consideration of how mental health professionals, artists, and critics collaboratively author mad history through engagement with performance—and what responsibilities might emerge from that collaboration.

Throughout, I use the term "mad" to reflect my alignment with critical mad studies, a discipline that "takes social, relational, identity-based, and anti-oppression approaches to questions of mental/psychological/behavioural difference."[2] At present, such differences are commonly articulated through the language of *mental illness*, framing mental disability and distress as biomedical diseases requiring diagnosis and treatment. While vital for some, this framework too often serves to discursively sever mental illness from madness, narrating the former as the apolitical subject of the sciences and the latter as the romanticized subject of the arts. In contrast, I follow mad activists and critical mad studies scholars who use the terms *mad* and *madness* to approach mental disability and distress "not only as medical conditions but also as historical formations that have justified all manner of ill-treatment and disenfranchisement."[3] In sum, madness, as a critical framework, positions "mental illness" as a historically specific concept, one that problematically privileges contemporary psychiatric theories and retroactively asserts their validity in the past. In doing so, it makes those concepts available for critique, positioning mad history and mad historiography as sites of political struggle.

Moreover, when applied to the Steppenwolf revival, it also invites consideration of performance's role within that struggle. Mad studies' interest in madness as a historical formation echoes performance studies' interest in *performativity*, a term that describes the way identity is constituted through embodied behaviors and their subsequent description—what theorist Judith Butler describes as "a stylized repetition of acts."[4] In other words, just as ideology is perceptible in text, so too is it perceptible in bodies—their gestures, vocalizations, and facial expressions indexing cultural attitudes. Thus, while historical work in mad studies has largely based its analysis on written documents, attention to performativity and performance may offer the discipline a sharper vocabulary through which to stage its key claims. When approached as historiography— as a form of historical writing—both performance-making and performance criticism, with their shared attention to the performing body, have the capacity to render apparent the performativity of madness while also dramatizing how that performativity has been mediated through critical acts in the clinic, the theatre, and beyond—often to ableist ends.

One Flew Over the Cuckoo's Nest is particularly well-suited to this task. First published as a 1962 novel by author Ken Kesey, adapted into a 1963 Broadway comedy by playwright Dale Wasserman, and transformed into a 1975 Academy Award-winning film by director Miloš Forman, *Cuckoo's Nest* remains a robust cultural artifact, appearing regularly on reading lists, sustaining steady theatrical revivals, and, most recently, inspiring a Netflix prequel series. The story follows rough-and-tumble Randle P. "Mac" McMurphy as he feigns insanity in order to avoid incarceration on a prison farm. However, upon his admission to a psychiatric hospital, he soon learns that he has inadvertently transformed his fixed prison sentence into an indefinite medical stay. Only the tyrannical nurse who rules the roost can approve his discharge—and she's convinced he really is ill. As her abuses increase, Mac leads his fellow patient-inmates in a revolt; in response, he is subjected to electroconvulsive therapy and, later, a lobotomy, rendering him unresponsive. Upon witnessing the damage done to his friend, Chief Bromden—an Indigenous patient-inmate and Mac's closest confidante—smothers Mac to death before smashing a window open and escaping into the night.

Scholars have largely interpreted the story's appeal within the broader counterculture of the 1960s, analyzing *Cuckoo's Nest*'s self-proclaimed anti-authoritarian philosophy alongside its problematic racist, sexist, and ableist politics. While important, less consideration

Table 1 One Flew Over the Cuckoo's Nest, *a Timeline of Adaptations*

1962	1975	
Novel Ken Kesey publishes *One Flew Over the Cuckoo's Nest*.	Film *Cuckoo's Nest* premieres in cinemas, winning Academy Awards for Best Film, Best Director (Miloš Forman), Best Screenplay (Bo Goldman), Best Actress (Louise Fletcher), and Best Actor (Jack Nicholson).	
1963		2000
Broadway Play Dale Wasserman's theatrical adaptation opens and runs for 82 performances.		Steppenwolf Revival The Steppenwolf revival runs for 10 weeks, followed by tours to London and New York the following year.
pre-1973: trauma-based theories	post-1973: neurobiological theories	

has been given to the plot's notable emphasis on performativity; it is Mac's enactment of mental illness that sets the action in motion. Performances of *Cuckoo's Nest*, therefore, raise important questions around what constitutes madness in a particular time and place and how that madness is discernible through a distinct performativity. Staged thirty-eight years after the original novel, thirty-seven years after the Broadway play, and twenty-five years after the feature film, the Steppenwolf revival provides an opportunity to consider how that performativity has transformed over time and how artists, critics, and mental health professionals have contributed to that transformation (Table 1).[5]

The remainder of this chapter proceeds in two parts. I begin with a dramaturgical analysis of the clinic, underscoring the theatricality of psychiatric spaces while pointing toward an emergent mad performativity produced, in part, by mental health professionals. I then consider how this performativity has shifted significantly in the years between *Cuckoo's Nest*'s original publication (1963) and revival (2000), noting that psychiatry in both periods produced distinct theories of madness with contrasting models of mad performativity. Given that actors reproduce aspects of this performativity on stage, I then consider how critics responded to the anachronistic performativity embodied by actors, providing historical context while adjudicating the accuracy of each response. By positioning these responses as a form of historiography, I demonstrate that critics not only narrated the Steppenwolf performances within the development of psychiatric theory, they also implicitly forged a history of mad performativity that has been collaboratively crafted by both clinic and stage. This work bears on the lives of mad people; therefore, I close with a consideration of how an understanding of criticism as historiography might point toward an understanding of criticism as care and the responsibilities associated with caring for mad pasts. I begin in the clinic.

Mad Performativity, Mad Performance

While a number of scholars have detailed the imbrication between medicine and performance, the general population does not typically understand the clinic as a performance venue. Yet, performance is central to its operations. For example, there is, at the time of this chapter's writing, no material test—no blood panel, no X-ray, no brain scan—available to indicate the presence of a mental disorder. Rather, in order to make a diagnosis and prescribe treatment, mental health professionals rely, in part, on their observations of a patient's performativity, interpreting gestures, movements, and facial expressions, among others, much like performance critics evaluate actors in a play—all while producing clinical records that are, in effect, medicalized performance reviews.[6]

These interpretations are further supported by criteria laid out in the American Psychiatric Association's (APA) diagnostic manual, *The Diagnostic and Statistical Manual of Mental Disorders (DSM)*, currently in its fifth edition. This 947-page tome details 157 diagnoses, each accompanied by corresponding criteria that, when observed in a patient, may indicate the presence of a mental disorder. While it would be reductive to equate the *DSM* with psychiatry writ large, the theatricality of these diagnostic metrics is striking and warrants further consideration. Take, for instance, a very small sampling of the current criteria on offer, each corresponding to one or more diagnosis: "disorganized speech,"[7] "prolongations of consonants as well as vowels,"[8] "repetitive, seemingly driven, and apparently purposeless motor behavior,"[9] "often runs about or climbs in situations where it is inappropriate,"[10] and "psychomotor agitation or retardation nearly every day,"[11] among many, many others.

Articulation, repetition, rhythm, sequence, locomotion—these are vocabularies of performers, of performance critics. Thus, for the purposes of this chapter, I propose that while the *DSM* presents itself as a diagnostic manual, it might be more productively approached as an aesthetic treatise, one that details intricate categories and subcategories of mad performativity identifiable through embodied expression. Taken together, these acts of clinical performance criticism suggest that theatrical logics infuse clinical practice, producing interdisciplinary methods that are part science, part performance.

To be clear, this framework does not in any way suggest that experiences of mental disability and distress are somehow not real; to the contrary, attention to mad performativity provides language to locate and name operations of power not easily apprehended in dominant medical paradigms, demonstrating that madness is not simply described but produced through its description. As mental health professionals undertake this descriptive work, they not only document their patients' performativity, they also determine what and how that performativity signifies. Thus, while vocal color, athletic physicality, patterned gestures, and so on may be narrated as skillful expression, when performed by mad bodies, these enactments are more often narrated as signs of pathology produced not by a sentient actor but, rather, by deficiencies in internal functioning caused by, for example, a chemical imbalance (i.e., too little serotonin, too much dopamine). This interpretive frame severely limits the expressive capacity of mad bodies performing in both formal and quotidian settings, centering mental health professionals as the primary arbiters of madness and its meanings while silencing the perspectives of mad people.

Despite this, there have been, and continue to be, significant changes in how mental health professionals theorize madness. Just as aesthetic preferences shift on the stage, so too do they shift in the clinic. Disorders come in and out of vogue, their theatrical diagnostic criteria expanded and contracted, adding a certain temporal instability to mental illness and its referent. These contingencies impact theatrical representations of madness, broadly, and the Steppenwolf revival, specifically. Psychiatry underwent a major change in the latter half of the twentieth century, a change that I understand as characterized by shifting attitudes toward mad performativity. In 1968, DSM-II identified 182 mental disorders; later, DSM-III (1980) identified 265, followed by 294 identified in DSM-IV (1994). What diagnoses do or do not appear is, in part, a function of shifting cultural attitudes. For example, prior to 1973, the *DSM* listed homosexuality as a mental disorder, a designation that was later removed from the manual's third edition. Medical historians often narrate this change as an example of when psychiatry misapprehended its subject, mistaking sexual orientation for mental illness. However, this change also initiated a reclassification of both mad performativity and gender performativity, as earlier diagnostic criteria problematically took perceived gender nonconforming behavior as indicative of improper sexual development. Thus, mad performativity is contextual; there is no performance of madness, formal or quotidian, that is not historically specific.

Moreover—and most consequentially for the Steppenwolf revival, I think—these changes were accompanied by a shift in how mental health professionals understood the mechanism of madness. Prior to 1980, the APA largely articulated mental illness as a subconscious disorder produced by unprocessed trauma. However, beginning in the 1960s, a growing anti-psychiatry movement posed a challenge to this model, accusing psychiatry of pathologizing nonconformist behavior while weaponizing medicine as a form of social control. While not explicitly named as such, theatricality was key to these critiques, positioning psychiatrists as faux doctors engaged in acts of mimetic malpractice. It was a charge to which mental health professionals were particularly sensitive, and in 1973—the same year homosexuality was removed from the *DSM*—the APA pivoted away from trauma theory, focusing instead on madness as a biomedical disease caused by physical differences in the brain.[12] Broadly speaking, this transition from trauma theory to a neurobiological model moved clinical models progressively inward, toward an increasingly molecular understanding of madness that continues to the present day.

The Steppenwolf production of *Cuckoo's Nest* straddles two sides of this temporal divide. The text, written in 1963 and lightly reworked in 1970, is saturated with earlier psychiatric diagnoses. For example, the script heavily implies that one character, Dale Harding, is gay, suggesting that his sexuality is the cause of his hospitalization. Wasserman's notes for the script include definitions for "Jocasta Complex," "Orestes Complex," "Oral Neurosis," and different, dated forms of schizophrenia such as "Reactive Schizophrenia" and "Process Schizophrenia."[13] These now-defunct diagnoses likewise cite earlier psychiatric theories, presenting characters whose madness stems from lingering trauma caused by overbearing mothers and oppressive social norms, a framework embodied by the play's overbearing and oppressive Big Nurse. And while Wasserman's script hardly takes these frameworks at face value, by the time the Steppenwolf revival opened in 2000, audiences well-accustomed to the brain chemistry of later theories would likely already find these diagnoses and models suspect on scientific grounds, their familiarity rigorously cultivated through direct-to-consumer marketing campaigns advertising the latest psychiatric medications.

Director Terry Kinney attempted to negotiate this representational tension by framing the 2000 revival as a faithful rendition of the 1970 Wasserman script, essentially approaching *Cuckoo's Nest* as a period piece. If some aspects of the text appeared dated, this was a feature, not a flaw, offering audiences an opportunity to reflect on "the ways in which our collective understanding of mental health and its treatment has changed over the last 35 years."[14] However, this set a particularly complex challenge for the production's actors, many of whom played mad characters. Realist scripts like Wasserman's require a particular performance technique, one in which the actor's outward behavior provides some clue to their internal motivation, positing a direct causal relationship between inner worlds and outer expression. Thus, when a grinning Mac declares to Nurse Ratched that he'll comply with her demands, we understand his grin, not his words, to reveal his true intent. His behavior communicates meaning. However, as theatre scholar Anna Harpin argues, the perception of madness in a character raises the possibility of a break in that causation; something about the inner-outer relationship has gone awry, making the character's true intent inaccessible. "Realism tends to remainder the contents of 'mad' experience in some ways outside the dramatic frame," Harpin writes, observing that "Madness in such works is, frequently, reduced to identifiable surface behaviors that are framed as 'ill,' behaviors that exceed the limits of the internal logic of the play-world."[15] When applied to mad characters, behavior too often communicates no meaning—only madness. The *Cuckoo's Nest* actors were, therefore, tasked with using acting techniques appropriate to a realist script while simultaneously using those same techniques to embody characters positioned beyond realism's representational mode.

In order to navigate this challenge, *Cuckoo's Nest* actors undertook extensive research, partaking in immersive site visits to two psychiatric institutions where, as one reporter put it, "the cast mixed with real psychiatric patients and nurses,"[16] before engaging in day-long improvisational group therapy exercises in character.[17] Actors also spoke to mental health professionals and patients, modeling their performances off of diagnostic criteria and ethnographic accounts.[18] This method-based research certainly allowed actors to develop more specific character choices; however, they remained limited by the script's largely realist frame.[19] Moreover, recalling the historical contingencies listed earlier, I want to suggest that actors likely studied a form of madness quite different from the madness portrayed in Wasserman's script. The resulting performances were, therefore, anachronistic, presenting audiences with a performance pastiche that merged elements of madness as it was understood in 1965 and madness as it was understood in 2000.

Critics, then, had to narrate these competing elements, identifying which madness—the madness of 1965 or the madness of 2000—actors were embodying in a given performance. For critics like Heidi Weiss, the difference between the two periods was so great that the play no longer signified madness. "Things change, of course," she observed. "So now you may watch 'Cuckoo's Nest' more as a simple human interest story."[20] For others, it was an unfamiliar, if not offensive madness, constituting a "ludicrous misunderstanding of psychiatric practice and its willful, all-but-criminal skewing of psychiatric theory."[21] In sum, the historical contingencies that impacted the Steppenwolf revival required that critics function as historians and historiographers, diagnosing each actor's embodied representation of madness and evaluating that representation's relevance to their current moment. As critics expressed preference for one style or another, they implicitly argued for one style of mad performativity over another, effectively merging clinical and critical discourse.

Critical Response

Across reviews, critics agreed that Wasserman's script left much to be desired, particularly in terms of its representation of madness. "It's rather late in the day to be chortling at the odd antics of the mentally disturbed," wrote Charles Isherwood, "but to approach the material with one's reason and sensitivity barometer intact is futile,"[22] arguing, however problematically, that the strength of the actors' performance transcended any critique of the text. However, for other critics, it was precisely the performers' "bravado acting"[23] that warranted critique. Their concerns revolved around the two shifts in mad performativity described in the previous section: changes in diagnoses and a shift from trauma-based models to a neurobiological model of mental illness.

Critics frustrated by actors' performances tended to argue that their externalized, physical choices too readily collapsed into exaggerated character types that were not simply over the top but problematically cartoonish. In his review, Joel Henning accused director Terry Kinney of allowing actors "to present an assortment of overdone and silly mental-patient tics" that would have been on "the edge of incredibility even in the 1960s."[24] This is particularly noticeable in the play's group therapy scenes, set in the wide, shallow dayroom of a psychiatric hospital. Here, Kinney crafted well-composed linear stage pictures, playing with shape and levels while deploying physicality to render each character's individual madness immediately legible. The result was a production that capitalized on the expressive capacities of the body in order to signify the inner workings of the mind, presenting audiences with outstretched legs, rounded shoulders, and folded arms, all skillfully posed in near-tableaux.

In these scenes, actors "settle on a single, identifying shtick and then repeat it with little or no variation for two hours,"[25] wrote critic David Finkle, an observation echoed by Ben Brantley of *The New York Times*. "Count 'em off: the stutterer, the twitcher, the screamer, the catatonic. They're embodied by an appealing group of performers," Brantley argued. "Probably too appealing, since they're as comfy a coterie of eccentrics as the barflies of 'Cheers.'"[26] While grounded in a critique of the production's directorial and acting practices, Brantley's assessment and those of his colleagues also reflect changes in psychiatric theory. In these reviews, critics regularly identify a dominant performance trait associated with each character. This is particularly noticeable in Brantley's use of a definite article—*the* stutterer, *the* screamer, *the* catatonic— underscoring the limiting nature of these enactments. However, by 2000, psychiatry had developed a far more robust theory of madness, one that included 297 distinct mental disorders each delineated by a complex combination of performative traits. The resulting performances do not seem to correspond to psychiatric theory so much as they rely on the representational idioms of madness; for this reason, the resulting acting choices appeared both impenetrable and undiagnosable, producing a production that appeared less mad and more random.

Other critics interpreted the acting style within the framework of trauma-based responses to emasculation, the kind that typified psychiatric theory at the time of the play's original production but had (at least officially) fallen out of favor by the Steppenwolf revival. In her review, critic Rhoda Koenig wrote, "mothers, it seems, have caused all the trouble,"[27] ultimately resulting in improper sexual development. Thus, when actor Ross Lehman "unleashes a big hysterical aria," or Danton Stone "collapses in weeping,"[28] the actors are not representing mental illness per se but rather an exaggerated gendered performativity historically pathologized as madness—a lingering association the revival showed little interest in critiquing. Characters' singular performative traits were not, in some critics' estimation, mad as much as they were trauma-based, and there was little if nothing to be found in these performances that suggested an absence of serotonin or a presence of misfiring neurons.

However, not all aspects of the production fall into this style, and it is here that an interesting tension emerges. Several reviews pay particular attention to the production's opening moments, which used lighting, sound, and projections to dramatize the hallucinations of Chief Bromden, diagnosed with schizophrenia. In this scene, lights flashed arrhythmically as heavy metal music blasted through the speakers, overlaid with the sound of roaring waterfalls, creaking gears, and a persistent electrical hum that later recurs in the play's electroconvulsive therapy scene. Actor Tim Sampson stood center stage in hospital fatigues, his body framed in a tight spotlight as he delivered a monologue to his characters' deceased father, describing, through metaphor, the inner workings of the hospital.

For numerous critics, this moment effectively staged the forces that "push at the edges"[29] of the character's mind, interpellating audiences into the somatic experience of schizophrenia and its symptomatology while cutting through the "realistic treacle" of the production.[30] However, these moments were few and far between, recurring only between key scenes throughout the production. In this sense, the incorporation of a more expressionistic framing served to weaken the other scenes by contrast, resulting in a production that was, as critics argued, "disappointingly sane"[31] and "not crazy enough."[32] Moreover, that this device was applied to the only patient-inmate of color is unfortunate, adding a further layer of othering to a character already shaped by racist stereotypes.

Despite this, echoing their dislike of earlier group scenes, critics' preference for these interludes similarly reflected changes in psychiatric theory. While the physical choices of the ensemble actors did not provide audiences with any clues as to their characters' internal functioning, Steppenwolf's *Cuckoo's Nest* scaled up and spatialized that internal functioning. Notably, in contrast to other scenes, the actor's body was almost unexpressive, holding stillness amid the swirling sensorium. In these moments, the production seemed to displace the character's madness from the actor's body, as if the latter could no longer signify the delicate choreography of neurochemicals partnering with neurotransmitters. And so, the stage design magnified and abstracted the brain, rendering its mechanisms perceptible through scenography.

When read alongside one another, critics' debate over the use of external, physicalized, ensemble-based scenes in contrast to preference for internalized, spatialized, individual moments in *Cuckoo's Nest* reflects a shift in psychiatric theory toward an increasingly molecular understanding of madness, one in which its mechanisms are not located in an amorphous subconscious but, rather, in a material neuroscience. Subsequent theatrical representations of madness throughout the aughts would follow a similar trajectory as Steppenwolf, using dramaturgy and design to effectively turn the mind inside out and expose its inner workings. Three months after the Steppenwolf revival opened, Sarah Kane's *4.48 Psychosis* premiered posthumously at the Royal Court Theatre, representing madness through a series of postdramatic scenes played out on a mirrored set. Anthony Nielsen's 2007 *The Wonderful World of Dissocia* used magical realism to represent dissociation through fantastical characters and settings. *Next to Normal* (2008) likewise made use of projections while casting an actor to play the main character's deceased son, allowing audiences to perceive her internal reality.

This move toward dramatizing a character's internal experience of madness lets audiences in on the proposed cause of their behavior, suggesting that madness has reason submerged within. However, whereas in earlier productions, that madness could be apprehended through theatrical performance, the critical response to *Cuckoo's Nest* suggests that this method of apprehension may no longer have been understood as reliable at the time of the revival. Indeed, the skillful stillness of Sampson's characterization has an almost filmic quality to it, echoing the tight close-ups focused on the eyes in the 1975 film adaptation's electroconvulsive therapy scene or, perhaps in inverse, *Next to Normal*'s use of large eyes projected, Gatsby-like, on two giant screens hanging above the stage. What I suggest, then, is that critics here are narrating an increasingly smaller, less embodied vocabulary of mad performativity, one that not only reflects the historical development of psychiatric theory but also perceives that history to be written on and through the performing body.

Critical Care

While clinic and stage are often considered separate domains, this analysis demonstrates that performance is central to our understanding of madness, now and in the past. Both clinic and stage share a porous boundary, one in which mental health professionals routinely rely on theatrical methods, engaging in acts of clinical performance criticism in order to produce a complex set of aesthetic categories pathologized as mad performativity. This performativity shifts over time, placing particular pressures on performers embodying madness on stage and performance critics responding to those representations. As the Steppenwolf reviews demonstrate, oftentimes these performances are adjudicated on clinical terms, focusing on whether or not a given performance reproduced a clinically sound embodied account of mad performativity and its underlying mechanisms. Couching their evaluation in a discussion of acting practices, arts critics not only rely on clinical frameworks, they also join in conversation with mental health professionals who serve as expert witnesses. By way of closing, then, I'd like to briefly consider the power dynamics of these conversations and their implications for future critical practice.

Here, I return to the article cited at the opening of this chapter. In addition to offering her own critiques, Keller's review also features quotes from several mental health professionals who offer their evaluations of the play and subsequent revival. For example, psychotherapist Mark Smaller observed, "The fact is, there is nothing funny or affirming about paranoia, hallucinations, or a breakdown. This is a very scary business." Susanne M. Andriukaitis, then executive director of the Chicago Chapter of the National Alliance for the Mentally Ill, echoed Smaller's assertion: "We now know that mental illnesses are not people's fault. For people who suffer from these illnesses, they affect every aspect of their lives." Summarizing Andriukaitis's comments, Keller noted, "The play is not just dated she believes, *it's downright dangerous*."[33]

It is laudable that these critics and mental health professionals articulated the pervasive impact representations of madness have on those experiencing mental disability and distress. However, as this chapter demonstrates, it would be a mistake to understand the work mental health professionals do in the clinic as vastly different from the work critics do in the theatre. If, as Steppenwolf asserts, the *Cuckoo's Nest* revival offers audiences a chance to understand how attitudes have changed over the years, then attention to mad performativity on the stage can help us to better understand how that performativity and, most importantly, the theories and frameworks through which we read that performativity are themselves historical formations. It might help us to better perceive how madness is constructed, what pressures have impacted its construction, and how that construction has been narrated across time. Performance criticism, then, might serve as a form of historiography, one that narrates, with nuance, the profoundly corporeal nature of so-called disorders routinely confined to a disembodied mind.

Approaching performance criticism as historiography might also prompt us to look beyond clinical accounts and toward other expressions of madness and its meaning in pursuit of anti-ableist critical practice. Analysis of critics' response to the Steppenwolf revival indicates that, at its worst, performance criticism can reproduce the most painful, depoliticizing aspects of the clinical encounter. In the reviews considered here, this is most apparent in critics' use of clinical frameworks coupled with a tendency to position those frameworks as ahistorical and, therefore, beyond the realm of politics.

At its best, performance criticism can destabilize clinical frameworks, providing new ways of thinking with madness while cultivating an attitude of epistemic generosity in response to mental disability and distress. This approach is most clear in reviews where critics explicitly narrate the changing aesthetics of clinical and cultural spheres, underscoring the historical contingencies that inform madness. For this reason, we might also understand artists and critics as located within economies of care, as caretakers of a particular history. A more intentional approach to criticism as historiography would not only account for historical contingencies but also take seriously criticism's own role in that history. Rather than accepting neat divisions between clinic and stage or drawing distinctions between mental illness as a medical disease and madness as an artistic device, critics might approach the two as interrelated, both equally informing the performativity of madness and the various resources and consequences that performativity generates in its particular time and place. Most importantly, such criticism might seek to be accountable to mad communities. It might understand that engagement with mad culture requires cultural competency. Finally, in the spirit of the disability mantra "nothing about us without us,"[34] an approach to criticism as care must center the perspectives of mad people, not only as characters but also as critics in their own right.

Notes

1 Julia Keller, "Psychological Drama 'Cuckoo's Nest' Is Dated—but Is It Dangerous?," *Chicago Tribune*, 12 May 2000, 5.
2 Rachel Gorman and Brenda LeFrançois, "Mad Studies," in *Routledge International Handbook of Critical Mental Health*, ed. Bruce M. Z. Cohen (London: Routledge, 2019), 107.
3 Tanja Aho, Liat Ben-Moshe, and Leon J. Hilton, "Mad Futures: Affect/Theory/Violence," *American Quarterly* 69, no. 2 (2017): 293.
4 Judith Butler, "Performative Acts and Gender Constitution: An Essay in Phenomenology and Feminist Theory," *Theatre Journal* 40, no. 4 (1988): 519. Other scholars have applied Butler to the study of mental/psychological/behavioral difference. While my analysis differs in its focus, my thinking is, nonetheless, shaped by their work. For more, consult Ellen Samuels, "Critical Divides: Judith Butler's Body Theory and the Question of Disability," in *Feminist Disability Studies*, ed. Kim Q. Hall (Bloomington Indiana University

Press, 2011), 48–65; Joy Venn, "Theatres of Mental Health," in *The Cambridge Companion to Theatre and Science*, ed. Kirsten Shepherd-Barr (London: Cambridge 2020), 116–30; Nick Walker, *Neuroqueer Heresies: Notes on the Neurodiversity Paradigm, Autistic Empowerment, and Postnormal Possibilities* (Fort Worth: Autonomous Press, 2021), 168–91.

5 Mental health professionals have been involved with *Cuckoo's Nest* from its inception, partnering with writers, actors, and directors to craft the story and its enactment. Additionally, the 1975 film continues to feature prominently in psychiatric trade publications and university courses. For more on mental health professionals' contributions to *Cuckoo's Nest*, consult Jennifer Lambe, "Memory Politics: Psychiatric Critique, Cultural Protest, and One Flew Over the Cuckoo's Nest," *Literature and Medicine* 37, no. 2 (2019): 298–324. For more on *Cuckoo's Nest* in publication and pedagogy, consult Deborah J. Boschini and Norman L. Keltner, "Different Generations Review One Flew Over the Cuckoo's Nest," *Perspectives in Psychiatric Care* 45, no. 1 (2009): 75–9; Jim Pink and Lionel Jacobson, "One Flew Over the Cuckoo's Nest," *BMJ* 334, no. 7594 (2007): 641; H. Steven Moffic, "We Are Still Flying Over the Cuckoo's Nest," *Psychiatric Times* 31, no. 7 (2014): 40.

6 Meredith Conti makes a similar observation in her analysis of nineteenth-century performance, writing that within Victorian medical science, "external symptoms of illness were expressions of benign or malignant changes to the body's anatomical structures." Conti's work extends Kirsten Shepherd-Barr's contention that spectators of nineteenth-century theatre functioned as amateur clinicians, evaluating an actor's performances to identify characters' illnesses. While Shepherd-Barr describes this process as a kind of Foucauldian diagnostic gaze, both Shepherd-Barr and Conti's analyses are particularly useful in that they move beyond ocularcentric paradigms toward a sharper consideration of embodiment and affect. For more, consult Meredith Conti, *Playing Sick: Performances of Illness in the Age of Victorian Medicine* (London: Routledge, 2018), 3; Kristen Shepherd-Barr, "The Diagnostic Gaze: Nineteenth-Century Contexts for Medicine and Performance," in *Performance and the Medical Body*, ed. Alex Mermikides (London: Bloomsbury Methuen Drama, 2016), 37–50.

7 American Psychiatric Association, ed., *Diagnostic and Statistical Manual of Mental Disorders: DSM-5* (Arlington, VA: American Psychiatric Association, 2013), 94.

8 Ibid., 45.

9 Ibid., 77.
10 Ibid., 60.
11 Ibid., 161.
12 For an excellent overview of this history, consult Decker, *The Making of DSM-III: A Diagnostic Manual's Conquest of American Psychiatry* (Athens, OH: Ohio University Press, 2013).
13 Dale Wasserman, "Definitions," t-mss 1996-027, box 23, folder 3, Dale Wasserman Papers, New York Public Library, New York, NY.
14 Martha Lavey, "Artistic Director's Preview," *Backstage News Magazine*, 2000.
15 Anna Harpin, "Dislocated: Metaphors of Madness in British Theatre," in *Performance, Madness and Psychiatry: Isolated Acts*, ed. Anna Harpin and Juliet Foster (New York: Palgrave Macmillan, 2014), 189.
16 Lawrence Bommer, "Amy Morton Feathers Her Cuckoo's Nest," *Theatre Mania*, April 3, 2000, theatermania.com.
17 While marketing for the Steppenwolf production would have readers believe that the actors' visit to a psychiatric facility was novel, such immersions are, in fact, well-worn, both in terms of *Cuckoo's Nest* and in terms of performance writ large. For more on *Cuckoo's Nest*, consult Charles Kisleyak, *Completely Cukoo*, directed by Charles Kisleyak (1997; Boston Quest Productions, 1997) DVD. For more on performance, consult Conti, *Playing Sick*; Jonathan W. Marshall, *Performing Neurology: The Dramaturgy of Dr. Jean-Martin Charcot* (New York: Palgrave Macmillan, 2016). Importantly, artists with lived experience of mental disability and distress have offered alternate methods for engaging with these kinds of site visits. For more, consult Petra Kuppers, Stephanie Heit, April Sizemore-Barber, V. K. Preston, Andy Hickey, and Andrew Wille, "Mad Methodologies and Community Performance: The Asylum Project at Bedlam," *Theatre Topics* 26, no. 2 (2016): 221–37; Petra Kuppers, Stephanie Heit, Lanxing Fu, and Jeremy Pickard, "Performance Artists Roundtable," in *Identity, Culture, and the Science Performance. Volume 1: From the Lab to the Streets*, ed. Vivian Appler and Meredith Conti (London: Bloomsbury, 2022), 17–33.
18 Fauzia Arain, "Johner Soars in 'Cuckoo's Nest,'" *The DePaulia*, May 5, 2000; Albert Williams, "The Actor's Grind," *Chicago Reader*, May 25, 2000; Emily Jenkins, "Untying the Knots," *Daily News*, April 8, 2001.
19 Here, I refer to what is commonly described as "method acting," developed by Lee Strasberg. A key component of this technique relies on "affective memory," wherein actors draw on personal memories

to create emotion and character. I describe Steppenwolf's work as "method-based" because its founders have, at various times, resisted the method label. For more, consult Patrick Marmion, "Why Terry's Still Mad for It," *Evening Standard*, 31 July 2000.

20 Heidi Weiss, "One Flew Over the Cuckoo's Nest," *The Chicago Sun-Times*, 17 April 2000.
21 David Finkle, "One Flew Over the Cuckoo's Nest," *Theater Mania* (blog), 9 April 2001.
22 Charles Isherwood, "One Flew Over the Cuckoo's Nest," *Variety*, 8 April 2001.
23 Simon Saltzman, "'One Flew Over the Cuckoo's Nest,'" *US1*, 1 May 2001.
24 Joel Henning, "This 'Cuckoo' Just Won't Fly," *Wall Street Journal*, 15 May 2000.
25 Finkle, "One Flew Over the Cuckoo's Nest."
26 Ben Brantley, "You're a Bad, Bad Boy and Nurse Is Going to Punish You," *The New York Times*, 9 April 2001.
27 Rhoda Koenig, "The Rise of the Half-Baked," *The Independent*, 31 July 2000.
28 Richard Christiansen, "Steppenwolf Resurrects 'Cuckoo's Nest,' but Is It Worth the Effort?," *The Chicago Tribune*, 17 April 2000.
29 Finkle, "One Flew Over the Cuckoo's Nest."
30 Chris Jones, "One Flew Over the Cuckoo's Nest," *Variety*, 24 April 2000.
31 Brantley, "You're a Bad, Bad Boy."
32 Ty Burr, "One Flew Over the Cuckoo's Nest," *Entertainment Weekly*, 17 March 2000.
33 Keller, "Psychological Drama 'Cuckoo's Nest' Is Dated," 5.
34 For more on this slogan, consult Michael Bérubé, "Representation," in *Keywords for Disability Studies*, ed. Rachel Adams, Benjamin Reiss, and David Serlin (New York: New York University Press, 2015), 151–4.

2

Through Fish Eyes

Raising Awareness of Ocean Degradation through Performance

Kasi V. Aysola and Madhvi J. Venkatesh

When we stand at the edge of a shore and look onto the horizon, we often have an emotional experience—perhaps a feeling of gratitude or wonderment at the magnitude of how small human existence is in comparison to the body of the ocean. Through their explorations and ongoing discoveries of new organisms and lifeforms, researchers continue to remind us that there is so much of the ocean that remains unknown. Scientists have done extensive work to understand how various marine species function within habitats and larger ecosystems, and yet we still attribute marine species' modes of living mostly to instinct and survival. What about emotion?

Scientists have done extensive work to understand the social structures of marine life, yet humankind often behaves in ways that assume that the emotional capabilities of humans are superior to those of other species. Don't animals feel? Are humans the only species that function based on complex social nuances? Are concepts like friendship and love exclusive to humankind? In fact, there are many marine animals that have strong family bonds and demonstrate the abilities to effectively communicate and work together.[1] Some of these social behaviors, such as the romantic courtship of whales, appear to be manifestations of emotion:

> Würsig (2000) has described courtship in southern right whales off Peninsula Valdis, Argentina. While courting, Aphro (female) and Butch (male) continuously touched flippers, began a slow caressing motion with them, rolled towards each other, briefly locked both sets of flippers as in a hug, and then rolled back up, lying side-by-side. They then swam off, side-by-side, touching, surfacing, and diving in unison. Würsig followed Butch and Aphro for about an hour, during which they continued their tight travel. Würsig believes that Aphro and Butch became powerfully attracted to each other, and had at least a feeling of "after-glow" as they swam off. He asks, could this not be leviathan love? [2]

Like these whales, many animals engage in courtship and mating activities and "seem to fall in love with one another just as humans do."[3] In fact, famed neuroscientist and psychobiologist Jaak Panksepp, who specialized in studying the neural mechanisms of emotions, asserts that love and other emotions are controlled by similar neural pathways in both humans and animals. He and his colleagues explain that "many species of warm-blooded vertebrates [have] a variety of basic emotional networks [that] are anatomically situated in similar brain regions, and these networks serve remarkably similar functions."[4] Thus, several other animals can experience romantic love and a variety of other emotions that humans feel through very similar biological processes.

We can observe animal behaviors that appear to be displays of emotional experiences. For example, a whale may carry the body of a deceased family member for days or even weeks in a seeming ritual of grieving. Sea lions express intense territorialism and competition for mates. Renowned ecologist and author Marc Bekoff explains that "current research provides compelling evidence that at least some animals likely feel a full range of emotions, including fear, joy, happiness, shame, embarrassment, resentment, jealousy, rage, anger, love, pleasure, compassion, respect, relief, disgust, sadness, despair, and grief."[5] The fact that some organisms can experience a gamut of "human" emotions provides an avenue for allowing people to connect with and better understand the experiences of animals.

The idea of empathizing with marine life inspired us as dancers and dance-makers who were trying to create an evocative piece. We felt that imagining the emotional intricacies of animals could be an effective way to motivate audiences to change behaviors. This concept continued to evolve and became the crux of our evening-length dance production, *Through Fish Eyes* (2019), which captured marine animals' complex range of emotions by humanizing these creatures and embodying their cognitive experiences. Through creating and performing this work, our ultimate goal was to deliver a call to action and raise ocean awareness. The lines of human and nonhuman animals were blurred in the performance as an attempt to dismantle the perceived hierarchy of humans over nature and the "crude dualism that put the human species on one side and all other [species] on another side."[6] Using dance, we sought to help audience members develop empathy and establish a sense of personal connection with marine lifeforms. In essence, we wanted to produce change, "not only in the ways that we live with animals and the ways we think about them, but also by transforming our values more broadly, resetting our priorities, [and] rebooting our sense of what it might mean to be human."[7] Like environmental humanities scholar Una Chaudhuri, we believe that "animal acts . . . are a powerful way to change the world."[8]

Humanizing the Ocean

Through Fish Eyes is a dance production that evokes empathy for dwindling marine ecosystems by confronting humankind's harmful relationship with nature. We conceptualized the dance production, which was performed by Prakriti Dance,[9] a Washington DC-based dance company specializing in Indian performing traditions. *Through Fish Eyes* premiered at the John F. Kennedy Center for the Performing Arts in May 2019 and has since toured across the United States (Rasa Festival, Ann Arbor, in June 2019; Battery Dance Festival, New York, in August 2021; University of South Florida, Tampa, in October 2021; Jacob's Pillow, Beckett, in July 2022; Rose Wagner Performing Arts Center, Salt Lake City, in September 2022; Dance Bloc, New York, in November 2022; Atlas Performing Arts Center, Washington, DC, in March 2023). The movement vocabulary for the piece is rooted in the Indian performing tradition Bharata Natyam, which fundamentally strives to express narratives through body language and facial expressions. Extensive training in Bharata Natyam has helped us understand how rhythmic dance and nonverbal storytelling can bring characters to life based on the accompanying music.[10] In both solo and group pieces, Bharata Natyam can convey the nuances of all types of stories and life experiences. In the context of *Through Fish Eyes*, our dance form equipped us with a rich framework to create impactful imagery and portrayals of marine life.

Our goal in this piece was to help bridge the perceived divide between humans and animals, which often supersedes humankind's phylogenetic ties to several other mammalian species.[11] Writer and researcher Melanie Challenger explains our fundamental similarities to other animals as follows:

> We're composed of cells with genetic material, and we move around, seeking energy to feed our bodies, pooping it out again as waste. We look a lot like our fellow primates with our five-digit hands and feet, our thoughtful eyes, and our lean, muscular physiques. We have lungs, a heart, a brain, a nervous system, and all those other features we share with mammals …. A reminder, if we still need one, that all life is our kin.[12]

Despite our close biological relationships to other animals, we have managed to isolate and distinguish ourselves from them, citing "our exceptional minds, capable of moral thought and free will."[13] This perspective has "allowed [us] to monopolise the Earth's resources and use other species for our own ends not because we're animals but because we are beings with special, unique properties."[14] We believe this disconnection from other animals, and the natural world more generally, along with humankind's sense of supremacy has led people to a sense of entitlement to natural resources. Since empathy involves taking on another's thoughts and feelings, we felt that it was an ideal tool for helping people understand the harm that their actions are causing to the animal world and for challenging the sense of human superiority and separateness that promotes environmental destruction. We sought to foster empathy by using narrative vignettes to help audiences understand the perspectives of marine animals and feel more personally connected to the experiences of lifeforms enduring the harmful effects of human behaviors.

Through Fish Eyes follows the relationship between humankind and the ocean, starting in a time when people lived in harmony with nature and ending with the current fragmented world. It gives a voice to water in its various forms by bringing to life the once abundant and thriving ocean and embodying its distress now that it has been ravaged by humans. It concludes with a call to action that urges humankind to change their destructive behavior.

The visual and musical design for the dance follows the trajectory of the production. The first half of the piece is more colorful and vibrant in all aspects. It reflects a more classical approach to the soundscape using the Carnatic music style from South India and *ragas* or melodies chosen to celebrate the wonder and magnificence of marine life.[15] In this section, saris are draped on the dancers' bodies and reflect this sentiment through their rich, deep ocean tones.[16] Suddenly, the second half of the piece takes a contrasting tone that reflects the current fragmented and artificial world. The soundscape takes an industrial turn incorporating sounds from factories and machines. Inspired by plastic mannequins, the costuming is stripped down to a transparent tunic, which reveals the contours of dancers' bodies and allows the audience to see the plastic shapes being created.

One of the dances created to evoke empathy was inspired by a scene from *Blue Planet II: Episode 4* showing a grieving whale mother after losing her calf. It was narrated as follows:

> A mother is holding her newborn young. It's dead. She is reluctant to let it go and has been carrying it around for many days. In top predators like these, industrial chemicals can build up to lethal levels and plastic could be part of the problem. As plastic breaks down, it combines with these other pollutants that are consumed by vast numbers of marine creatures. It's

possible her calf might have been poisoned by her own contaminated milk. Pilot whales have big brains. They can certainly experience emotions. Judging from the behavior of the adults, the loss of the infant has affected the entire family. Unless the flow of plastics and industrial pollution into the world's oceans is reduced, marine life will be poisoned by them for many centuries to come.[17]

This grieving process of whales was so visceral that we decided to embody it in the piece. Rather than a literal representation, we personified the mother-child relationship and infused a human thought process into the characters of a whale and calf to evoke the most relatable response. The scene is prefaced with a narration by a small child, personifying the whale calf, narrating how it is caught in the net of a fisherman. The whale calf says, "I can't get out, *Amma* (Mom)! I can't get out!" This narration is a grim foreshadow of what is to come in the scene.

The scene begins with the mother whale waking up her sleeping baby as it is time to begin their day. The calf gets up reluctantly (as most sleepy children do) and begins to swim and play. Then comes time for a feeding from the calf's mother (Figure 2). After filling its belly, the calf scurries away mischievously while the mother whale runs behind. The mother warns the calf not to cross a certain boundary, but the calf does not heed her warning. As the calf swims further away from the mother whale, it unknowingly enters the net of a greedy fisherman.

Then, a musical shift begins, foreshadowing the danger ahead for the calf. The sole male dancer enters swinging a net prop, representing the violent hands of humankind, and captures the calf in the net. The mother whale realizes what is happening and frantically tugs at the net, trying to free her baby with no success. The calf is dragged away and the lighting becomes red, symbolizing death. The improvised violin solo echoes the sounds of a crying whale, and the mother shrieks in pain after losing her child. The dance then transitions into an abstract movement sequence fusing the ideas of human loss and the mental anguish of an animal.[18] This is by far the most visceral and painful scene of the entire performance, often bringing tears to the eyes of the audience.

FIGURE 2 *Mother whale feeding her calf (dancers from left to right: Ramya Kapadia and Archana Raja; photo credit: Jay Pillai of Rudram Dance Company).*

Dancing at the Boundaries of Science, Art, and Culture

Through Fish Eyes was created by observing ocean life as inspiration to communicate through dance. We tested the hypothesis that dance could serve as a powerful embodied approach to build empathy for life beyond the common realm of sight through work-in-progress showings in multiple cities across the United States. Bharata Natyam is uniquely positioned in the mainstream world of dance because of its cultural ties and the exoticized lens through which it is often viewed.[19] Even though Bharata Natyam may be viewed as culturally specific,[20] its versatile movement vocabulary has allowed it to extend far beyond its Indian cultural origin. As the form continues to evolve, many dancers and choreographers have presented works that deal with social and environmental issues and stories from around the world. Through our years of Bharata Natyam training and performance experience, we were well-versed in the range of narratives and messages that the dance style could communicate. Therefore, we were confident that it would be an effective medium for a production that artistically translated the intricacies and dynamics of marine ecosystems into movement.

To create *Through Fish Eyes*, we blended artistic and scientific approaches by engaging in observation and communication, which are essential for both endeavors. On one hand, we used an extensive observation process, iterative testing during development, and synthesis into a full-length communicative piece, which are elements of "the real process of science."[21] On the other hand, we focused on aesthetics and creating emotional responses, which are often considered a part of the artistic process.

As explained in literature on learning and development, observations need to be connected to context and culture to go beyond merely seeing.[22] This mental process allows us to understand what we are seeing by linking it to existing knowledge and experience. Observation can inspire a work of art and shape its perspective, regardless of whether it takes a representational or abstract approach.[23] Representational approaches aim to reproduce or mimic the subject in a way that is recognizable. Abstraction veers away from representation by trying to bring about the nonvisual essence of the subject without directly showing it. *Through Fish Eyes* uses the integrative approach of weaving together iconic representational imagery and abstraction to evoke emotional responses.

The representational aspects of *Through Fish Eyes*, which ranged from coral reefs to a turtle injured by a plastic straw, showcased an embodied understanding of nonhuman entities generated by observation. For instance, we physicalized a coral reef landscape through dancing bodies inspired by hours of snorkeling, visiting aquariums, and watching video footage. The segment mimics the intricacies of coral reef ecosystems, including hard and soft corals that are inhabited by various creatures. A cluster of bodies created the overarching structure of the reef, with dancers' extremities functioning as soft corals moving with the ocean's currents. Buried within, and occasionally emerging into plain view, were a variety of fish, crabs, sea stars, eels, clams, and seahorses, represented by the hand gestures of dancers crouching within the reef structure. Ethereal music and shades of blue and green in the lighting and

FIGURE 3 *Coral reef (dancers from left to right: Kasi Aysola, Vanita Todkar, Madhvi Venkatesh, Seema Viswanath, Madhavi Reddi, Ramya Kapadia; photo credit: Siva Sottallu).*

costumes accented the embodied reef by emulating a buoyant underwater experience (Figure 3).[24]

In addition to using representations of both the magnificent beauty of the ocean and the terrible harm that humankind has caused to marine life, we leveraged abstract representations of the ocean to create a call to change. In showing the metaphorical beating down of the ocean and its subsequent rising in anger against humankind, we sought to convey a sense of urgency in improving the plight of the ocean. The final scene of the piece shows a male character, representative of all humans, manipulating and harming a female character, depicting the ocean and marine life. As this happens, other female dancers at the back of the stage slowly emulate rising water levels to signify the ocean's building sense of anger and frustration.[25] Eventually, the ocean's energy can no longer be contained and all those representing the ocean unite in a series of tumultuous waves that encircle and subsume the male character representing humankind. As the water settles, the audience hears voices calling for change in an effort to harness the emotional intensity of the scene (Figure 4).[26]

FIGURE 4 *Ocean brought to life in anger from "Hands of Change" (dancers from left to right: Vanita Todkar, Kasi Aysola, Madhavi Reddi, Madhvi Venkatesh, Seema Viswanath, Ramya Kapadia; photo credit: Siva Sottallu).*

Defining the Scope of the Performance

It was not always clear that *Through Fish Eyes* could be so heavily dependent on direct observation. The project was initially focused on the mesopelagic zone of the ocean, also known as the "Twilight Zone." This zone extends about approximately 200–1,000 meters deep and has little to no sunlight penetration.[27] The mesopelagic was a specific focus of our initial funder, the National Academies Keck *Futures Initiative* (NAKFI). They wanted people to better understand this part of the ocean; it has incredible diversity but is relatively unexplored. Like all parts of the ocean, the mesopelagic is negatively impacted by humans, but much of the impact is still unknown. As we began researching and developing *Through Fish Eyes*, our hope was to support NAKFI's vision by creating an art-science dance production about the mesopelagic that would bring awareness and visibility to this unknown part of the ocean.

After working for about a year, we started to realize that embodying a relatively unknown zone of the ocean may not offer a visceral experience for viewers. Questioning our original intentions, we recognized that we did not just want to bring forth something new; we desired to elicit change in human actions that had a detrimental impact on the ocean. This message of social change was not limited to one specific part of the ocean; therefore, we needed to move beyond interpreting facts about the mesopelagic region.[28]

We began to think about the concept of change itself. Change is often caused by necessity or triggered by tragedy. Experiencing trauma or hardship forces humankind to grow and improve. We wanted to prompt empathetic responses in our audiences with the hope that they would then assess how their lifestyles affected the ocean and environment. Similar to how Una Chaudhuri describes another artist's work, we wanted to use storytelling and visuals "to offer a kind of somatic knowledge, a way of understanding the Other by going beyond rationalizations . . . to physicalization."[29]

This also necessitated broadening the piece's scope to include familiar stories and creatures that would allow audience members to relate to the characters more easily. We wanted to personify marine animals in a way that tapped into a familiar visual lexicon of ocean life by embodying them, for example, as friendly dolphins or nurturing whales. We also sought to use vivid imagery to illustrate the wonder of the ocean, such as intricate seashores and vibrant coral reefs, as a contrast to gruesome scenes of dying sea turtles, murdered whales, and the graceless, extractive practices of humans.

To make the scenes as affective as possible, we brought in iconic, image-driven stories from social media. *National Geographic*'s Jane J. Lee recounts one such story about a team of scientists that spent almost ten minutes pulling a single-use plastic drinking straw from the nostril of an olive ridley sea turtle. The team, which included sea turtle expert Christine Figgener, helped the injured reptile off the coast of Costa Rica. Figgener thought that the turtle swallowed the straw, gagged on it, and then tried to throw it back up. After the incident, Figgener said, "We couldn't believe what we had just pulled out of that turtle . . . Usually, trash such as plastic bags and even toothbrushes end up in a sea turtle's stomach. It's also quite common to see fishing hooks embedded in a turtle's mouth or flipper."[30] This story made us realize how many human-created objects harm animals and that our piece must capture the magnitude of humankind's impact on marine life.

Inspired by the viral video of Figgener's team helping the injured sea turtle, we developed a section of the dance about this story to engage with audiences' preexisting knowledge of how single-use plastic products can harm marine life. The dance sequence begins with a plastic straw visible in a spotlight, emphasizing its critical role in the scene as an object of harm. Soon after, the male human character enters the spotlight and retrieves a drink from a cooler on a hot day. The human happily enjoys the drink using the plastic straw, completely oblivious to any potential negative consequences. In the meantime, a sea turtle collectively formed by four dancers comes into view and slowly approaches the center of the stage. Using deliberate movements, the four dancers move together as a composite turtle consisting of a large shell with four legs, a tail, and a head extending out of it. Once the size of the turtle has been established, the dancer forming the front portion of the animal remains in the role of the turtle while the remaining three dancers emulate being strangled by a noose, ominously foreshadowing the turtle's fate.

By this time, the human character has finished his drink. After looking around to make sure that no one is looking at him, he discards the straw. In his hands, the straw then makes its way to the ocean (like the 5.25 trillion pieces of human-generated trash currently in the ocean)[31] and is carried by waves through whirlpools and eddies. Eventually, it reaches the unsuspecting turtle that has been happily exploring. The power of the ocean currents painfully lodges the straw in the turtle's nose, and it is visibly in discomfort as it attempts to remove the foreign object. As the turtle struggles, the human returns and seeing the turtle in pain, runs toward it and tries to relieve it by removing the straw. The scared turtle attempts to retreat from the human's hands, but the human eventually pulls out the straw after a bit of struggle. As the human extracts the straw and holds it up, the turtle immediately falls dead, fatally injured by the straw's insertion and its subsequent removal.[32] This scene highlights the vulnerability of these large, seemingly strong creatures to the single-use plastics that humans often thoughtlessly use and discard.

Communicating Unfamiliar Concepts through Movement

It took iteration and experimentation to make the performance's abstract concepts as communicative and impactful as the familiar representational scenes like the turtle and the straw. This was particularly true when embodying objects like plastic that are foreign to the Bharata Natyam dance vocabulary, which consists of a complex system of gesture and movement but is traditionally tied to an Indian cultural context. The first version of our performance expanded the conventional uses of the Bharata Natyam movement vocabulary by showing the progression of plastic from its molecular components to the objects that it makes. The dancers embodied the polymerization of individual atoms into chains that then combine together to make the moldable plastic material that humans use to construct various objects, including fishing nets and plastic straws. Using tetris-like patterns of movements and the shapes of the dancers' bodies, the piece then showed the physical weaving of a net and the creation of a straw, both objects of violence and pain in later parts of the production.

The second iteration of the performance reenvisions this section to emphasize the creation of a virtually indestructible material like plastic, rather than focusing on showing the material's progression into objects of violence. In fact, plastic was originally created to protect the environment by replacing ivory, thereby reducing the mass slaughter of wild elephants.[33] Ironically, this synthetic material, intended to reduce unnecessary animal killings and overuse of natural resources, has become a highly lethal substance for nature.

In the second iteration, one dancer plays the role of a "mad scientist" (inspired by the inventor of synthetic plastic, Leo Baekeland)[34] with the other dancers moving in a slow, amorphous unit symbolic of a malleable substance that could be stretched and shaped by the mad scientist. The scientist then directs mass production of plastic straws through a line of dancers repeating the robotic movement patterns of an assembly line machine.[35] Unlike the first iteration where the entire cast showed the plastic material and its by-products, this revised version introduces a human character to personify for audiences the rapacious, controlling nature of humankind.

The two iterations are both innovative in their depiction of synthetic plastic, a material unfamiliar to the Bharata Natyam context, though they differ in their precise choreography. The first iteration heavily depended on the physical stance, *hastas* (gestural vocabulary), and footwork of the Bharata Natyam form to show the progression of plastic creation from its molecular components to human-made objects of harm. In contrast, the second iteration represents the fluidity of melted plastic and its molding into a solid form by relying on the dancers' collective improvisations rather than prescribed choreographic movements.

Leveraging Gender Identities to Chronicle the Human-Earth Connection

The piece leveraged the performers' gender identities in representing the evolving relationship between humankind and nature. In the first half of the piece, all the dancers in the cast are part of a collective whole, demonstrating the coexistence of humans and nature and evoking the splendor of marine life. In contrast, *Through Fish Eyes*' second half draws a clear distinction in roles between the sole male dancer and female dancers. This portion, inspired by plastic's chemical synthesis and its subsequent commercialization, begins with a male scientist conceptualizing an easily manipulatable material; controlling the shapeless plastic mass embodied by the female dancers; and ultimately directing the female dancers in the mass production of plastic straws. The piece then progresses to show the male dancer consciously and unconsciously harming the lives of marine animals—the straw-impaled sea turtle and the whale mother and calf brought to ruin by a fishing net—and ultimately the ocean as a whole. In both cases, the male character representing humankind is the perpetrator of harm, the antagonist of the plot. These vignettes are followed by an abstraction: the male dancer controls and abuses a female dancer who ultimately joins with the cast's other female dancers to outdo him. In a fundamental metaphor of *Through Fish Eyes*, the collective female dancers represent the rising anger of the ocean exacting retribution on an aggressive, often unrepentant humankind. The ocean (and its rage) ultimately cannot be contained or controlled, despite the best efforts of humankind. As we stress in *Through Fish Eyes*, a lack of harmony with the ocean—and nature more generally—will ultimately harm humans. In the words of Una Chaudhuri, "we have to be concerned about the other animals not only for their sakes but also for ours."[36]

The casting, which subsequently challenges existing gender hierarchies between men and women, serves as an allegory for broader social inequities that result in manipulation, conscious and unconscious harm, and sometimes retaliation. It illustrates that just as we should seek harmony and respect in our relationships with nature, we should strive to level the hierarchies between people with different identities. After all, social groups with fewer economic resources are often most reliant on their relationships with natural resources and at the greatest risk of harm from environmental destruction. Therefore, protecting these communities involves not only safeguarding the natural resources upon which they rely, but also empowering them and elevating their voices through greater social equality, thereby intertwining social justice and environmental conservation.

Shattering Human Hierarchies

We humans tend to view ourselves at the top of the food chain when in reality we are somewhere in the middle.[37] Our false sense of human superiority has forged ideologies and systems of extraction, overconsumption, and destruction, thereby disrupting nature's equilibrium. This perception of dominance also fosters a belief that nature lacks sophistication and nuance. In relation to the ocean, we often believe that marine organisms possess "predator and prey" or "eat or be eaten" mentalities; this misapprehension neglects the reality that these ecosystems are highly nuanced.

The ramifications of humans' perceived sense of superiority sparked our idea that if somehow the ocean and its life-forms could be humanized through dance, perhaps audiences would comprehend the dangerous impacts of their actions. This led us to imagine marine life through a human lens, dismantling the human-animal hierarchy by treating animals as humans and humans as the animals that they are. We infused emotion into a world that humans have deemed separate and less superior, staging an octopus's mental dialogue and imagining an orca's friendship with a fish. We brought the ocean to life through the bodies of the female dancers and allowed the "voice" of the ocean to speak against its human abuser. By humanizing these animals and the ocean, we leave viewers with the message that human-like empathy needs to be directed to marine organisms to drive change.

Acknowledgments

The project described in this book chapter was supported in part by the National Academies Keck *Futures Initiative* of the National Academy of Sciences under award number NAKFI DBS18. The content is solely the responsibility of the authors and does not necessarily represent the official views of the National Academies Keck *Futures Initiative* or the National Academy of Sciences. Additional support for this project was provided by grants from the Boston Foundation's Live Arts Boston program and the Boston Cultural Council.

We also gratefully acknowledge the contributions of all the artists who were a part of creating *Through Fish Eyes*. In particular, we thank the other members of Prakriti Dance who have performed in the various iterations of this production, Ramya Kapadia, Archana Raja, Madhavi Reddi, Madhuvanthi Sundararajan, Vanita Todkar, Seema Viswanath, Sophia Salingaros, Amrita Doshi, Ricardo Roman, and Priyanka Raghuraman, for the ways in which their ideas, movement, and energy influenced the development of the piece. We are also grateful for the musical contributions that Ramya Kapadia and Anjna Swaminathan made to this production.

Notes

1 "5 Marine Animals with Strong Family Bonds," *Aquaviews Online Scuba Magazine*, 24 May 2017, https://www.leisurepro.com/blog/explore-the-blue/5-marine-animals-strong-family-bonds/.

2 Marc Bekoff, "Animal Emotions: Exploring Passionate Natures: Current Interdisciplinary Research Provides Compelling Evidence that Many Animals Experience such Emotions as Joy, Fear, Love, Despair, and Grief—We Are Not Alone," *BioScience* 50, no. 10 (2000): 861–70, https://academic.oup.com/bioscience/article/50/10/861/233998.

3 Ibid.

4 Jaak Panksepp and Lucy Biven, *The Archaeology of Mind: Neuroevolutionary Origins of Human Emotions* (New York: W.W. Norton & Company, 2012), 1-2.

5 Bekoff, "Animal Emotions."
6 Una Chaudhuri, "Animal Acts for Changing Times, 2.0: A Field Guide to Interspecies Performance," in *Animal Acts: Performing Species Today*, ed. Una Chaudhuri and Holly Hughes (Ann Arbor University of Michigan Press, 2014), 2.
7 Chaudhuri and Hughes, *Animal Acts*, 1.
8 Ibid.
9 Prakriti Dance is an innovative dance company that was founded in 2014 by Kasi Aysola and Madhvi Venkatesh. Currently directed by Kasi Aysola, it takes the movement vocabulary of yesteryears and interprets modern-day themes, bringing relevance and context to the ever-evolving Indian performing arts. Drawing inspirations from nature, philosophy, poetry, and other genres of art, Prakriti weaves a multilayered tapestry to transcend cultural boundaries and communicate the human experience. The company specializes in presenting Indian performing traditions, and all company members have received extensive training in these art forms from leading exponents. The company has performed internationally at various prestigious venues including the Kennedy Center (Washington, DC) and Jacob's Pillow (Massachusetts). Prakriti Dance's work has been supported by several organizations including the National Academies Keck Futures Initiative, Maryland State Arts Council, the Boston Foundation, the Boston Cultural Council, the Arts and Humanities Council of Montgomery County, and the Barr Foundation. More information about the company and its dancers can be found at www.prakritidance.com.
10 Kasi and Madhvi both trained in Bharata Natyam under Guru Viji Prakash for over ten years. Kasi has also studied and worked with Mythili Prakash and other leading exponents of Indian dance. As a versatile and composite artist, Kasi has trained in other Indian performing traditions, including Kuchipudi dance and Carnatic music. In addition to her training under Guru Viji Prakash, Madhvi has refined her understanding of Bharata Natyam under the guidance of esteemed stalwarts including Guru T. K. Kalyanasundaram, Harikrishna Kalyanasundaram, Indira Kadambi, and Sreelatha Vinod. Both Kasi and Madhvi have performed solo and group work in the United States, India, and abroad and have received recognition and funding for their artistic work from various local, state, and national organizations.
11 Morris Goodman, John Czelusniak, Scott Page, and Carla M. Meireles, "Where DNA Sequences Place *Homo sapiens* in a

Phylogenetic Classification of Primates," in *Humanity from African Naissance to Coming Millennia*, ed. Philip V. Tobias, Michael A. Raath, Jacopo Moggi-Cecchi, and Gerald A. Doyle (Florence: Firenze University Press, 2001), 279–89.

12 Melanie Challenger, "Are Humans Animals?" *BBC Science Focus Magazine*, 25 February 2021, https://www.sciencefocus.com/the-human-body/are-humans-animals/.

13 Ibid.

14 Ibid.

15 Carnatic music originates from the south of India and layers the components *rāgas* (melodic scales), *tālas* (time cycles), and *sāhithya* (lyrics or poetry). These components are improvised upon by the musicans, which is known as *manodharma*. Instruments that can be found in Carnatic music include veena, flute, mridangam, tambura, nagaswaram, thavil, violin, mandolin, and saxophone.

16 Prakriti Dance, *Through Fish Eyes: Water Cycle*, Prakriti Dance, 2022, Video, prakritidance.com/gallery.html. https://youtu.be/Qrz324Tctr8.

17 *Blue Planet II: Episode 4*, David Attenborough, BBC Studios, 2017, Documentary.

18 *Through Fish Eyes: Whale's Lament*, Prakriti Dance, 2021, Dance Video, prakritidance.com/gallery.html.

19 Rohini Acharya and Eric Kaufman, "Turns of 'fate': Jack Cole, Jazz and Bharata Natyam in Diasporic Translation," *Studies in Musical Theatre* 13, no. 1 (2019): 9–21, https://doi.org/10.1386/smt.13.1.9_1.

20 Janet O'Shea, *At Home in the World: Bharata Natyam on the Global Stage* (Middletown, CT: Wesleyan University Press, 2007), 3.

21 "The *real* Process of Science," *Understanding Science*. University of California Museum of Paleontology, https://undsci.berkeley.edu/article/0_0_0/howscienceworks_02.

22 Catherine Eberbach and Kevin Crowley, "From Everyday to Scientific Observation: How Children Learn to Observe the Biologist's World," *Review of Educational Research* 79, no. 1 (2009): 39–68, https://doi.org/10.3102/0034654308325899.

23 Jose-Antonio Soriano-Colchero and Inmaculada López-Vílchez, "The Role of Perspective in the Contemporary Artistic Practice," *Cogent Arts & Humanities* 6, no. 1 (2019), https://doi.org/10.1080/23311983.2019.1614305.

24 *Through Fish Eyes: Coral Reef*, Prakriti Dance, 2022, Dance Video, prakritidance.com/gallery.html.
25 The denoted genders are based on the way that the dancers self-identify.
26 *Through Fish Eyes: Hands of Change*, Prakriti Dance, 2022. Dance Video, prakritidance.com/gallery.html.
27 Roland Proud, Martin J. Cox, and Andrew S. Bierley, "Biogeography of the Global Ocean's Mesopelagic Zone," *Current Biology* 27, no. 1 (2017): 113–19, https://doi.org/10.1016/j.cub.2016.11.003.
28 NAKFI gave us considerable freedom for our piece to evolve, given the exploratory and interdisciplinary nature of the project. They were fine with us choosing to depict marine life more broadly because it still fit within the proposed concept of using dance to raise awareness about ocean organisms and the impacts that humans are having on them.
29 Una Chaudhuri, "The Silence of the Polar Bears: Performing (Climate) Change in the Theater of Species," in *Readings in Performance and Ecology*, ed. Wendy Arons and Theresa J. May (New York: Palgrave Macmillan, 2012), 52.
30 Jane J. Lee, "How Did Sea Turtle Get a Straw Up Its Nose?" *National Geographic*, 17 August 2015, https://www.nationalgeographic.com/animals/article/150817-sea-turtles-olive-ridley-marine-debris-ocean-animals-science.
31 Sonam Chaturvedi, Bikarama Prasad Yadav, Nihal Anwar Siddiqui, and Sudhir Kumar Chaturvedi, "Mathematical Modelling and Analysis of Plastic Waste Pollution and Its Impact on the Ocean Surface," *Journal of Ocean Engineering and Science* 5, no. 2 (2020): 136–63, https://doi.org/10.1016/j.joes.2019.09.005.
32 *Through Fish Eyes: The Turtle and the Straw*, Prakriti Dance, 2021, Dance Video, prakritidance.com/gallery.html.
33 "History and Future of Plastics," *Science Matters: The Case of Plastics*. Science History Institute, https://www.sciencehistory.org/the-history-and-future-of-plastics.
34 Joris Mercelis, *Beyond Bakelite: Leo Baekeland and the Business of Science and Invention* (Cambridge, MA: MIT Press, 2020), 162.
35 *Through Fish Eyes: Polymer*, Prakriti Sance, 2022, Dance Video, prakritidance.com/gallery.html.
36 Chaudhuri and Hughes, *Animal Acts*, 2.

37 Sylvain Bonhammeau, Laurent Dubroca, Olivier Le Pape, Julien Barde, David M. Kaplan, Emmanuel Chassot, and Anne-Elise Nieblas, "Eating Up the World's Food Web and the Human Trophic Level," *Proceedings of the National Academies of Sciences* 110, no. 51 (2013): 20617–20, https://doi.org/10.1073/pnas.1305827110.

3

Laboring the Medical

Female Bodies for Sale on the Contemporary Stage

Gianna Bouchard

Theatre's entwinement with science has generated a multitude of ways of staging and examining different forms of scientific work. This ranges from plays about scientists, such as Anna Ziegler's *Photograph 51* (2008), which focused on Rosalind Franklin and her contribution to the discovery of the structure of DNA, to the devised work of Complicité in *A Disappearing Number* (2007), which staged the working out of a mathematical problem (of infinite series) in parallel with the real-life story of the collaboration between mathematicians G. H. Hardy and Srinivasa Ramanujan. In these instances, scientific work is aligned with the highly technical and skilled practices of the scientific community and who are represented laboring in laboratories and academic environments to generate new techniques and theories. These examples of theatrical representations of scientific labor involve traditional forms of laboratory and intellectual work, investigating chemicals and mathematical formulae, in ways that align with modes of scientific experimentation and knowledge production from the early modern period onward.

Since the end of the twentieth century, however, the focus of that labor has shifted, and science and medicine have become increasingly invested in the manipulation of life itself. This has been identified as a turn to "biomedicalization," from "enhanced control over external nature (i.e., the world around us) to the harnessing and transformation of internal nature (i.e., biological processes of human and non-human life forms)."[1] This change to science work that deals with the "wet" components of biology has likewise been reflected on the contemporary stage with plays that focus on the materiality of the body, its tissues, cells, and even its genetic potentiality. For instance, Ella Road's play *The Phlebotomist* (2018) examined a future where a single blood test determines your life chances through providing a social rating based on things like predisposition to disease and illness and heritable character traits. At the same time, performance artists have sought to embed themselves within the wet biology lab, for example, at the University of Western Australia's research facility SymbioticA, where artists engage experientially in laboratory work linked to the life sciences. Through direct engagements with tissue culturing (where cells are grown and manipulated outside of the original body), artists such as Kira O'Reilly and ORLAN have undertaken their own scientific work to raise bioethical questions about these techniques and their implications.[2]

Manipulating biological life creates value or, more specifically, "bio-value," a term biomedical scholar Catherine Waldby developed to define the application of technology to living organisms in a way that extracts value from them, including economic and intellectual value.[3] This work continues to be attributed to the intellectual and practical work of the scientist and not to the bodies or the individuals upon which any such innovations depend: the donors and other necessary collaborators in this labor, human and nonhuman. In this chapter, I will explore theatrical representations of these differently laboring bodies in biomedical contexts. How has theatre in the UK in the late 2010s performed those whose bodies labor temporarily in the clinic or the laboratory, under the supervision of medical professionals, corporations, and the biotech industries more broadly? What about those whose bodies are precious to biomedicine but whose "embodied productivity" is often overlooked by science, legal frameworks, and wider society?[4] Cooper and Waldby call this kind of scientific work "clinical labour" and identify women as a particularly valuable source because of their reproductive capacities. Laboring as oocyte (egg) donors, surrogates, and clinical test subjects, women have become the new precariat of biocapitalism. Referring to the processes involved in extracting bio-value from bodies and body parts, biocapital seeks innovation, transformation, and investment in global markets of exchange. This chapter argues that contemporary theatre has been contributing to debates around the ethical and social dilemmas facing women entangled within this global economy and staging their lived experiences as precarious laborers.

In my first case study, *Guinea Pigs on Trial* (2014–16) by Sh!t Theatre (UK), the two performers recounted their experiences as young women trying to earn money through participating in clinical trials. The devised show used a variety of means, such as song, satire, film, dance, and verbatim interviews to perform the politics and ethics of such experimentation. My next examples are Vivienne Franzmann's play *Bodies* and Satinder Chohan's *Made in India*, both from 2017, which interrogate the repercussions of international gestational surrogacy on the women involved. Each of these surface the ways that women in different geopolitical contexts become precarious laborers through their bodies under globalized biotechnological regimes.

Testing the Subject

Sh!t Theatre is comprised of queer, white performers Louise Mothersole and Rebecca Biscuit, who have been making small-scale, devised work together since 2010. Often taken from their own experiences of social and political issues, such as job seeking or unknowingly illegally subletting a council flat in London, their work is a collage of different elements of performance, including multimedia, comedy, song and dance, political commentary, audience participation, and satire, all staged through a feminist lens. In *Guinea Pigs on Trial* (2014–16), Mothersole and Biscuit recounted their real-life experiences of trying to earn money through participating in Phase 1 clinical trials as test subjects.[5] The first of four phases of medicinal trials in humans, Phase 1 usually evaluates safe dosage, side effects, and efficacy. A key part of the bioeconomy, these trials are used by pharmaceutical companies to develop new or repurpose existing drugs for novel markets. The show homed in on the exploitative forces underpinning the machinations of this economy and how trial participants, male and female, are made vulnerable by and through their clinical labor as test subjects.

Early in the show, the performers explained how and why they came to make the piece, coming across adverts describing clinical trials posted in an employment center where they were job seeking. They found these adverts "strange" and quickly deduced that they were targeting "poor people," who were being encouraged to have "experimental drugs tested on them."[6] Taking on the mantle of Agents Mulder and Scully from the television series *The X Files*, they decided to investigate.[7] Opting to go undercover, they explained that they wanted to explore why people apply to participate in clinical trials, what might be dangerous about these trials, why doctors sign nondisclosure agreements in relation to this work, and why the results of these trials are often hidden. So, they attempted to engage in clinical trial "guinea pigging," a term used by an "underground community to refer to people who make their living by participating in . . . paid medical trials."[8] Mothersole and Biscuit deliberately tried to sell their bodies, and *Guinea Pigs on Trial* was created from this research investigation, exposing some of the practices and processes used by commercial trial companies to harness specific forms of clinical labor.

As a form of temporary and precarious employment, the clinical trial recruit can earn money by permitting their body to be used for experimental research, labor that can be identified as "closely resembling other kinds of casualized service work and drawing in the uninsured [in the US context] and unemployed."[9] Through song, Mothersole and Biscuit skewer these dynamics as they narrate their attempts to join a trial, with some of the following lyrics:

> It's a zero regulation
> Un-protected occupation
> On a drug's trial
>
> [. . .] If you are unemployed
> And *The Daily Mail*'s annoyed
> Do a drug's trial

> If you're feeling poor
> There's a temporary cure
> On a drug's trial.[10]

Humorously, they highlight the ethically dubious and exploitative aspects of these practices. Gradually, the economics of these trials were defined more explicitly, when the audience learned that a trial Biscuit applied for was offering a fee of £3,400 ($4,633) or the equivalent of £200 ($272) per day. They clarified that UK regulations mean that trial fees must be kept "purposefully low so as not to act as an incentive to 'risky or degrading circumstances.'"[11] They let this hang in the air for the audience to ponder, as £200 per day is obviously an enticing sum if you are unemployed or seeking ways of supplementing an income. The performance made it clear that while it is an individual's choice to sign up as a clinical trial participant, such choice might be seriously circumscribed by an individual's social and economic circumstances. Reflecting on one trial relating to heroin substitutes, they noted that for the assembled would-be test subjects being screened there was "clearly a fiscal necessity for them being there."[12] In other words, it made for a troublingly unequal power relation between the clinical laborer and the research company.

The show itself was partly framed as a form of clinical trial, with Mothersole reading part of a clinical trial consent form aloud to the audience. Biscuit meanwhile encouraged the audience to reply together by calling out "I consent" after each line. The consent was playfully edited to relate directly to the show. For example, lines referred to the unknowability of outcomes for the audience, with Mothersole reading: "I understand the possible risks and side-effects of the performance being tested" and "[i]t has been explained to me that the performance may contain risks to me that are currently unknown."[13] After each line, the audience responded with a cheery "I consent." This corresponds with the experience of being on a Phase 1 clinical trial, where the participant is expected to take on any associated risk with the experiment through informed consent. In this sequence, the artists also emphasized some of the more invasive and perhaps unexpected requirements of a clinical trial, asking the audience to verify that they understood the "contraception requirements" and that they would follow "these measures," even though this was the first time that these have been mentioned.[14] This referenced the kind of labor expected of test subjects and their material impact on the biological self. There will often be requirements around compliance with certain regimes on a trial, be they dietary, pharmaceutical, or physiological, such as relating to sleep, exercise, drug taking, and so forth. This is often a "labor of ingestion and metabolic self-transformation" that relies on an unacknowledged form of collaboration between the clinical laborer and the clinical team overseeing the trial.[15] Setting up this piece of metatheatre, drawing parallels between the performance and clinical trials as sites of experimentation, risk, and unknowability, allowed the duo to playfully juxtapose the sometimes alarming aspects of the expectations and dangers placed on clinical trial participants with their own work as experimental performance artists.

In various ways, the show highlighted how individuals are recruited to clinical trials—or not, as Mothersole and Biscuit were ultimately rejected from most of the trials for which they applied. Using the eligibility criteria for a clinical trial from Richmond Pharmacology (a London-based company dedicated to carrying out patient trials), they encouraged the audience to further consider the kind of disciplined and highly prescribed labor expected. The audience were asked to all raise their hands and then, as they read out the eligibility list, lower them at the point when they were no longer suitable. The list ran from having to be a nonsmoker or within a particular age range or a specified body mass index to requiring the participants to exclude themselves from eating a list of exotic fruit for the trial's duration, including pomegranates, pomelos, and grapefruit. Slowly but surely the audience's arms were lowered, until only a smattering remained who were then selected by Mothersole and Biscuit for a faux trial, testing vodka against water: cue a light-hearted bit of audience participation involving a drinking game.

Alongside staging the kinds of eligibility criteria used for Phase 1 clinical trials, the duo revealed that elimination from a trial can be worrying for a would-be participant but that these anxieties are not addressed by the trial companies. For instance, Biscuit was rejected from a gastroenteritis trial because of "inappropriate" levels of bacteria in her stomach.[16] She had no idea whether this was a good or bad thing and the company offered no further information, with the performers noting that it's not worth the company's time or money to do any kind of follow-up for rejected applicants. Mothersole and Biscuit offered several examples of these negative outcomes, including being rejected from Flu Camp, paid blood donation, and an asthma trial; each time Biscuit expressed worries that the professionals left the reasons for her trial rejections unexplained. Such anxiety points toward the ways that test subjects' bodies are experimented on, put on the line for, and utilized by trial companies with very little protection or support.

Through performing how they tried to sell their bodies to biomedical science, Sh!t Theatre highlights the exploitation and precarity of *all bodies* in relation to participation in forms of intimate labor, including clinical trials. As this chapter seeks to evidence, theatre is becoming increasingly interested in such bioprecarity, which Griffin and Leibetseder define as "the vulnerabilization of people as *embodied* selves . . . created through regulations and norms that encourage or require individuals to seek or provide bodily interventions of different kinds, in particular in relation to intimacy and intimate labour."[17] Toward the end of their research and the show they specifically considered women's bodies in a section on "disease mongering."[18] This is a tactic used by the pharmaceutical industry, which "widen[s] the boundaries of treatable illness in order to expand markets for those who sell and deliver treatments."[19] The performers listed some recently identified female disorders, the drugs devised to treat them, their side effects, and some of the research used to validate the need for pharmaceutical interventions. They improvised a doctor-patient interaction based on marketing for a treatment for premenstrual dysphoric disorder (pMDD), rhetorically asking the difference between pMDD and the more familiar premenstrual syndrome and noting that one of them "requires expensive hardcore anti-depressants."[20] During the improvisation they explained that the "drug company Eli Lilly rebranded Prozac [an antidepressant] as Serafem" for the condition and that this was clearly targeted at women because "it has 'fem' in the name, and it is pink."[21]

As a result of not being accepted onto any trials, the pair realized that women often get rejected unless that they are postmenopausal or sterilized. Deciding to find trials for women, Mothersole applied for a trial testing Botox on women as a therapy for an overactive bladder—a trial that they sang about to the tune of *Lucille* by Kenny Rogers:

> I was kind of wary
> It looked sort of scary
> The method looked pretty extreme
> An injection of Botox
> Directly into the bladder
> To paralyse and stop the stream
>
> I find Botox unnerving
> Incredibly disturbing
> Paralysing with poison is vile
> But I'm willing to be Botoxed
> And I'm waiting with legs crossed
> to get money by getting on the trial.
>
> You picked a fine trial to aim for Louise
> If your bladder stops working you will not be pleased
> If they give you Botox
> So that the flow stops
> It might harm you permanently
> You picked a fine trial to aim for Louise.[22]

At the time of staging the work in 2014, Mothersole had still not received the results of her screening for this trial. But, finally, the audience was informed that Biscuit had been accepted into a trial to test an experimental diabetes drug and was paid £33 per day. They joked that this was match funding for the show and that the pay was equivalent to being an intern.

Given the significant political and social issues implicit in this form of clinical labor, Sh!t Theatre's performance, predicated on their attempts to sell their bodies into clinical trials, was part journalistic investigation and part lived experience, interwoven with references to *The X Files* and a metatheatrical narrative about the trials of making theatre and the hunt for ways of financing such a career—which might include being a test subject. In highlighting the bioprecarity of all bodies in the contemporary moment, this chapter will now turn to the more explicitly female form of clinical labor of cross-racial surrogacy. Through this example, I am interested in "precarious bodies that engage intimate labour, i.e. those who seek the help of others in relation to body work, and with those who do that labour."[23]

Wombs for Rent

My next two examples explore international surrogacy and women's bodies as a vital source of reproductive clinical labor. Vivienne Franzmann's play *Bodies* and Satinder Kaur Chohan's *Made in India* were both staged in 2017 in the UK and consider the ethical, social, and emotional stakes involved when motherhood becomes entwined with biocapital. In both plays, a white, middle-class woman is involved in paying for a brown woman to act as a gestational surrogate, brokered through fertility clinics based in India. Both plays, therefore, deal with the complex personal and political issues that emerge from medical and social forms of dependency and exploitation that cross international borders. They also choose to intersect with the implementation of a real surrogacy ban in India, which impacts on the trajectory of their narratives and the course of each surrogacy. The ban came into effect in India in 2015, and this real-world political decision became a means for both playwrights to further explore the ethics of international surrogacy and the intimate reasons that women seek gestational surrogacy.[24]

Surrogacy involves the implantation of an embryo into a woman who has agreed to carry the pregnancy. Before the implantation, the surrogate must be readied through an intensive hormonal and pharmaceutical regime. The eggs are harvested, assessed for viability, and then inseminated with sperm in a laboratory, in each case from the intended parent or from a donor. The fertilized and selected embryos are then implanted into the surrogate's uterus. Gestational surrogacy refers to instances where the oocytes are those of a donor or the intended mother, and so the baby has no genetic connection with the surrogate: "[i]n effect, gestational surrogacy separates women's reproductive capacities into three elements: genetic (egg donor), gestational (surrogate), and social (intended mother)."[25] The interrogation of these divisions is an important part of these plays, both in their embodiment on the stage and in their themes.

In Franzmann's *Bodies*, staged at the Royal Court Theatre, 43-year-old, white, British television producer Clem and her husband Josh have purchased eggs from a Russian vendor to be implanted into 23-year-old Lakshmi, an Indian surrogate who has two children of her own and who has moved to the fertility clinic in an Indian city for the duration of the pregnancy. The eggs will be fertilized with Josh's sperm, and the couple are paying

£22,000 for this arrangement. The play moves between scenes in the UK, at Clem's home and that of her father, David, and the clinic in India, while also bringing these locations together digitally through a number of Skype calls. The play stages the journey of the surrogacy arrangement, examining Clem's desire to be a mother as she goes through the process, while also dealing with her father's deteriorating health. Clem's father has motor neurone disease and needs constant care, delivered by agency staff in the figure of Oni, another woman of color. Oni's husband and daughter are "back at home," while her son is with her in the UK, in a clear parallel with Lakshmi's situation, of a woman undertaking precarious and affective labor, separated from her family.[26]

The play's first scene is between Clem and her teenage daughter, and it starts as a straightforward domestic and mundane moment in their lives. The familial and close ties between them are expressed through reference to their relatives, to the way that Clem recollects her daughter as a baby and notes that she has her father's toes. Franzmann establishes a strong sense of genetic kinship in these early exchanges, which then start to unfurl. The daughter wonders why some animals reject their offspring or invest time and effort in incubating their young, only to disappear at the earliest opportunity. Clem then asks which school her daughter attends and whether she's happy there. She asks, "[a]re you here?" and seems fearful of her daughter leaving or not being there, and the tables get turned, with the daughter comforting Clem and saying that she will "make up for all the others."[27] Gradually, the daughter is revealed as an imaginary projection of Clem's, a version of the daughter that she is hoping for through assisted reproductive technology (ART) and a way for Franzmann to further question Clem's motivations and the social expectations of motherhood, as they entangle with cross-racial surrogacy.

It is Oni, the care worker assigned to look after Clem's father in his own home, who notes the inclusion of a Russian woman's eggs in the pregnancy, while the gestational surrogate is Indian, and the sperm is coming directly from Josh, the intended father. She stumbles into the realization that Clem and Josh want to ensure that "the baby's skin will be . . . ," with "white" left unspoken at the end of her sentence. She clarifies a few lines later by stating, "[y]ou want her to look like you," in a move that seeks to alleviate Clem's obvious discomfort as the racialized aspects of her endeavor are highlighted.[28] White privilege, which circulates through both of these plays in the form of educated, middle-class, affluent women who can afford to pay for ART, is punctured by confrontation with the racialized women on whom, in various ways, they depend. Clem's reliance on a Russian egg vendor is not explored any further than this moment of realization about the genetics of gestational surrogacy, where Clem will have no biological connection with the child. In fact, Clem states that she knows nothing more about this donor and neither does the audience. This aspect of ART is being increasingly "outsourced" to other, poorer countries such as Russia, and oocyte donation is even less likely to be recognized as labor within clinical bioeconomies because of its occurrence within the woman's body.[29] It is, nonetheless, a risky and time-consuming form of biological labor that reproductive clinics rely on to be performed by cheap female laborers, a point that is similarly overlooked by Clem.

In Scene 5, Clem is with her imaginary daughter once more, talking about Clem's previous attempts to have a baby and the fifth pregnancy that nearly reached full term. The daughter starts to articulate the dreams that Clem had for the child, "[t]he one that was going to be a doctor or beauty therapist or a judo instructor" and that Clem "would have killed a city full of people to protect her."[30] Suddenly, she asks if the surrogate, Lakshmi, is feeling the same way about the new baby, whether she loves it and is fiercely protective of it. Clem replies: "No. You're not her baby."[31] This reaffirms the perceived split in ideas of motherhood between the separate clinical labor of the gestational surrogate and the claim of the intended mother as the source of maternal love and the rightful parent in social and legal terms.

This tension also surfaces in *Made in India*, where surrogacy complicates maternal claims and the audience witnesses the brutal assertion of the rights of the contracting mother over the surrogate, enforced by the fertility clinic's female director. The play was written by Satinder Kaur Chohan and commissioned by Tamasha Theatre Company, which was founded in 1989 and specifically promotes work by artists from the global majority. It was initially performed in 2017 at the Belgrade Theatre, Coventry, and stages Eva Roe's journey to surrogacy following the death of her husband and her lengthy legal battle to gain access to his stored sperm. Set in Gujarat, India, in and around the fertility clinic of Dr. Gupta,

Eva becomes increasingly involved with her surrogate, Aditi, a 28-year-old Indian woman from a rural village. Eva promises much to the clinic's director and Aditi as the Indian surrogacy ban, which threatens to derail the pregnancy and transfer of the baby, comes into effect. In her focus on the pregnancy, however, Eva neglects and eventually loses her job, jeopardizing the whole arrangement. Meanwhile, Aditi, who is pregnant with twins, loses one of them and runs back to her home in a rural village, where she gives birth to one dead and one living baby in a distressing final scene. In the epilogue, Eva takes the baby from Dr. Gupta as the clinic is closed, and the last words are those of Aditi, asking to be a surrogate again.

Kaur Chohan writes in her author's note about *Made in India* that the idea for the play came to her "after reading a shocking article in which a middle-class English woman described an Indian village surrogate as her 'vessel.'"[32] This idea of a womb for rent, or the surrogate's body as simply a container, suggests a kind of property market that deliberately negates the experiences of the surrogate and their contribution. As a biological container for hire, the pregnancy is viewed as a temporary occupancy that exemplifies the operations of biocapital—body parts for sale across international borders, formally contracted through often unregulated fertility clinics. On the surface, these contracts obfuscate the considerable biological and emotional labor of the participating women as buyers and sellers of reproductive capacities. This is borne out in the final scene, where Dr. Gupta arrives after Aditi has given birth to the two 'white' babies, one of whom she is holding, while the other is dead in a plastic bucket. As Aditi starts to breastfeed the child, Dr. Gupta immediately advises against it; it's not in the surrogacy contract. Aditi tries to stake her own claim on the baby, but the doctor describes her as "just a vessel" and that the baby is "made to order. She's paying."[33] The contract trumps any of Aditi's claims to the child, and she is treated as no more than an incubator or a kind of babysitter for the duration of the pregnancy. In fact, both plays indicate how these arrangements are made to "[shore] up a fictive consensus on the definition of the social roles of mother and father."[34] This tension in the expectations of a surrogate to be a committed mother during the pregnancy but then able to give the child to the intended parents after the birth is also highlighted in *Bodies*, when the imaginary daughter tells Clem that "you can't want her to love me and not love me as well."[35]

The payment for surrogacy in these cases heightens a sense of entitlement for both women to the bodies and clinical labors of their surrogates, a claim over Aditi and Lakshmi that the plays seek to explore. In *Made in India*, Eva arrives at her selected clinic at the start of the play and demands to choose and then meet her surrogate, Aditi, a situation that is usually prohibited. Their developing relationship is then staged through the rest of the play, as Eva's assumptions about Aditi unravel, as she discovers that Aditi's husband was killed by a drunk driver and that she has two daughters back in her village. At the same time, Aditi discovers that while Eva has provided her eggs for the surrogacy, the sperm comes from Eva's dead husband. Aditi is horrified to learn that she has a dead man's sperm inside her, as she puts it, and it adds to the strain that both women are under as the surrogacy arrangements are tested to the limit. In *Bodies*, Lakshmi is separated from Clem and Josh when they visit the clinic in India, as contact between surrogate and intended parents is not permitted or encouraged. To this end, the audience witness Lakshmi coming onto the stage, looking for a way into the clinic, with the stage directions noting that she "is waved away and told to enter a different way."[36] It's a moment of apparent separation for contractual reasons, but it's also a recognizable form of racial and class segregation underpinning the clinical labor being staged.

Ultimately, *Bodies* and *Made in India* depict continuing forms of colonialism, now transposed to biocapitalist practices. The social pressures, the financial and personal implications of what some call reproductive tourism on the women involved, and the rippling outwards of those choices to close family and kinship networks, the politics of the market, and its exploitative aspects are all interrogated by Franzmann and Kaur Chohan. The usually unmarked bodies of poor women from the Global South are shown as deeply entangled with those of the white women purchasing their fertility and other forms of affective labor. The repercussions of transnational surrogacy, the lines of dependency, intersectionality, and its fragility are all laid bare.

At their centers, both plays explore the ethical dynamics of a financial transaction that implies fair treatment and recompense for the surrogate's labor while obscuring the risks, racialized ideologies, and inequalities at stake in these cross-racial arrangements. Both white women initially believe that the financial contracting of a racialized surrogate is a positive choice for these brown women, allowing them to improve their lives and the lives of their families. This is problematized in both plays, as Eva and Clem are made to confront their racial privilege through understanding the lives of the surrogates, the reasons for their choosing to become surrogates, and the impact of those choices. In *Made in India*, for example, Aditi reveals that she has no other choices available to her to make such a large amount of money and that this decision has meant that she has been rejected by her own family, who believes the surrogacy arrangement to be "dirty."[37] *Bodies*' Clem discovers that Lakshmi is in fact a widow and has no one to look after her own children while she is staying in the clinic for the pregnancy. It is Clem's socialist father, however, who is most outspoken about the inequalities that Clem herself seems unable to consider. The father's sense of social justice is articulated, importantly, by his carer Oni, who speaks on his behalf as his health declines. It is Oni who states that "in India, the surrogate has no rights" and that "Lakshmi has no protection. No one is looking out for her."[38] David expresses shame at Clem's surrogacy arrangements, which she battles by trying to explain the devastation that she feels at not being a mother, that otherwise she has no purpose or social status.[39] Even though she is reminded once again that the baby will not be (genetically) hers, the symbolic value of social motherhood, personally and collectively, is made clear and serves to naturalize the cross-racial surrogacy.

The figure of Oni in *Bodies* aligns clinical labor, in the form of surrogacy, with other forms of feminized labor, such as domestic labor and care work. Laura Harrison notes that "[c]ross-racial gestational surrogacy in particular is situated within a historical continuum of racialized reproductive labor whereby the economically and racially dominant classes rely upon the embodied and affective labor of women of color."[40] This is a dependency that Clem finds very difficult to acknowledge and manage, as she witnesses Oni giving her father the love and attention that he requires and she fails to provide. The premiere of *Made in India* also referenced other forms of feminized labor in scenic designer Lydia Denno's mobile fabric screens. Used as a means of delineating the imagined spaces on the stage, they also suggested the Indian textile and garment industries. The work of caring, domestic labor, and making garments rely on the use of the female body and this is heightened or exacerbated in the case of scientific labor performed in these examples, with more direct biological interventions into the body, in risky and often troubling ways.

The plays' interventions into the politics and ethics of surrogacy are also made visible through their staging of these women in the same spaces. Usually, "money facilitates anonymous exchanges between strangers," where the women have no direct contact with each other, but in these works the writers imaginatively bring the women together in the same locations and at the same time.[41] In *Made in India*, this is facilitated by Eva spending most of the pregnancy at the clinic and making demands to spend time with Aditi, for all of its tragic consequences for Aditi. In *Bodies*, the surrogate Lakshmi ghosts Clem's home life in several scenes, sometimes unseen by Clem but visible to the audience. She offers reflections on the surrogacy and speaks through Clem's imaginary teenage daughter—a projection of the child that Clem is desperate to have and who unpicks the politics and underlying exploitative practices of Clem's contract with the Indian fertility clinic. In analyzing reproductive labor, Catherine Waldby clarifies that ART means that "embryos and oocytes can be disentangled from the female body and transferred between different subjects. The biology of human fertility is hence opened up to flexible spatial possibilities."[42] These transactional, biological, and maternal connections are embodied and performed in the plays, collapsing the usual distances between these women and revealing the complexities of selling reproductive labor.

Neither play offers any comfortable or straightforward answers to these complications. Eva gets her baby at the end of *Made in India* but leaves in her wake Aditi dealing with the death of one twin and Eva having taken the other. It's a graphic penultimate scene of the moments after the birth; at the very end of the play, the audience is left with an image of Aditi telling the clinic director that she needs "to go again," making herself available again for a surrogate pregnancy.[43] Franzmann's *Bodies* ends with overlapping images of Clem and Josh in a hotel room with their new baby, of Clem's father at home, sitting immobile in front of a television with a new carer, of Lakshmi outside the clinic, lactating while holding a bag of money, and an image of a young Indian girl dancing on a table.[44] In both cases, however, the white, wealthy heterosexual, nuclear family remains intact, as even its traces exist in Eva's inclusion of her deceased husband's sperm in the fertilization of the embryo. While the plays mark important steps in illuminating the complexities of cross-racial surrogacy and bringing this new form of clinical labor to the stage, the central heterosexual family unit is left undisturbed. For many, ART offers the possibility of disrupting these conventional familial ties and creating alternatives, but in these theatrical examples such avenues are not pursued.

In conclusion, all three examples in this chapter represent new forms of precarious labor, in clinical trials and in oocyte donation and gestational surrogacy, which are aligned with other feminized forms of affective and intimate labor. The female, including the racialized and "othered" body, is performed in these examples, illuminating the risks and dangers inherent in bioprecarious work. Ultimately, these works raise significant bioethical questions about the sale of the female body across international and corporate borders.

Notes

1 Adele E. Clarke, Janet K. Shim, Lauran Mamo, Jennifer Ruth Fosket, and Jennifer R. Fishman, "Biomedicalization: Technoscientific Transformations of Health, Illness and U.S. Biomedicine," *American Sociological Review* 68, no. 2 (April 2003): 164.

2 See Oron Catts and Ionat Zurr, "Artists Working with Life (Sciences) in Contestable Settings," *Interdisciplinary Science Reviews* 43, no. 1 (March 2018): 40–53; Harriet Curtis and Martin Hargreaves, *Kira O'Reilly: Untitled (Bodies)* (London: Intellect and Live Art Development Agency, 2018); Simon Donger, Simon Shepherd, and ORLAN, *ORLAN: A Hybrid Body of Artworks* (London and New York: Routledge, 2010).

3 Catherine Waldby, *The Visible Human Project: Informatic Bodies and Posthuman Medicine* (London and New York: Routledge, 2000), 33.

4 Melinda Cooper and Catherine Waldby, *Clinical Labor: Tissue Donors and Research Subjects in the Global Bioeconomy* (Durham and London: Duke University Press, 2014), 100.

5 I would like to thank Louise Mothersole and Rebecca Biscuit of Sh!t Theatre for generously providing me with a copy of their unpublished technical script for *Guinea Pigs on Trial* for the purpose of writing this chapter.

6 Louise Mothersole and Rebecca Biscuit, *Guinea Pigs on Trial* (unpublished technical script, 2014), 1–26, 4–5.
7 *The X Files* was an American science-fiction television series that aired from 1993 to 2002 and was very popular on both sides of the Atlantic. The series followed FBI agents Fox Mulder and Dana Scully as they investigated paranormal happenings.
8 Mothersole and Biscuit, *Guinea Pigs on Trial*, 6.
9 Catherine Waldby, "Medicine: The Ethics of Care, the Subject of Experiment," *Body & Society* 18, nos. 3 and 4 (2012): 185.
10 Mothersole and Biscuit, *Guinea Pigs on Trial*, 2–3.
11 Ibid., 15.
12 Ibid., 18.
13 Ibid., 1.
14 Ibid.
15 Cooper and Waldby, *Clinical Labor*, 135.
16 Mothersole and Biscuit, *Guinea Pigs on Trial*, 14.
17 Gabriele Griffin and Doris Leibetseder, "Introduction," in *Bodily Interventions and Intimate Labour: Understanding Bioprecarity*, ed. Gabriele Griffin and Doris Leibetseder (Manchester: Manchester University Press, 2020), 1.
18 Mothersole and Biscuit, *Guinea Pigs on Trial*, 9.
19 Ray Moynihan, Iona Heath, and David Henry, "Selling Sickness: The Pharmaceutical Industry and Disease Mongering," *BMJ (Clinical Research ed.)* 324 (2002): 886.
20 Mothersole and Biscuit, *Guinea Pigs on Trial*, 9.
21 Ibid., 9–10.
22 Ibid., 23–4.
23 Griffin and Leibetseder, "Introduction," 6.
24 See Radhica Ganapathy's essay in Volume 1 of *Identity, Culture, and the Science Performance* for more information on this ban, "Staging Science and Humanity in Manjula Padmanabhan's *Harvest*," 215–36.
25 Laura Harrison, *Brown Bodies, White Babies: The Politics of Cross-Racial Surrogacy* (New York: New York University Press, 2016), 3.
26 Vivienne Franzmann, *Bodies* (London: Nick Hern Books, 2017), 31.
27 Ibid., 10–11.
28 Ibid., 29–30.

29 Catherine Waldby and Melinda Cooper, "The Biopolitics of Reproduction: Post-Fordist Biotechnology and Women's Clinical Labour," *Australian Feminist Studies* 23, no. 55 (2008): 59.
30 Franzmann, *Bodies*, 36.
31 Ibid., 37.
32 Satinder Kaur Chohan, *Made in India* (London: Samuel French, 2016).
33 Ibid., 66.
34 Harrison, *Brown Bodies*, 29.
35 Franzmann, *Bodies*, 60.
36 Ibid., 18.
37 Chohan, *Made in India*, 33–4.
38 Franzmann, *Bodies*, 74.
39 Ibid., 75–6.
40 Harrison, *Brown Bodies*, 33.
41 Waldby, "Medicine: The Ethics of Care, the Subject of Experiment," 281.
42 Ibid., 268.
43 Chohan, *Made in India*, 71.
44 Franzmann, *Bodies*, 124.

Creative Interlude

Please Let Me Shoot You: a monologue

Claudia Barnett

Special thanks to following artists and advocates for their roles in the development of this monologue: Vivian Appler, Emma Dreher, David Ian Lee, and Karen Sternberg (Pipeline-Collective).

Character

Odessa   *Small but fierce. Waterproof clothes.*

Setting

A Zodiac on the Saint Lawrence River at dawn.

Note

This scene was inspired by "Zodiacs and Crossbows: I Spent a Day Chasing Whales to Learn How to Study Them," by Justin Taus (vice.com, 1 September 2016).

   At rise. A small woman aims a large crossbow as she scans the ocean (the audience). Suddenly, a humpback whale emerges from the water in front of her, splashing. (SHE sees it; we don't.)

Odessa

Penelope! You gorgeous girl. I'd know that blowhole anywhere.

> *Water sprays, the Zodiac sways, and* **Odessa** *stumbles.*

Whoa! That's close.

> **Odessa** *scrambles to retrieve the crossbow.*

O-kay. Just—stay—still.

> SHE *aims.*

This isn't gonna hurt. I promise.

> *The whale is gone.*

No, wait!

> *The water is calm.*

Penelope? Don't you trust me?

> SHE *scans the water desperately, ready to shoot.*

I can see where the crossbow might look threatening, but it's not a weapon. It's a means of communication.

I have questions. You have answers. You may not think you know the answers, but they're in you. And all you have to do is let me shoot you. A tiny patch of cells can tell me what you've eaten and where you've been as well as information about your social structure, reproduction, and even the genealogy of your cetacean species.

 SHE *indicates the arrow in the crossbow.*

You see this teeny tiny, arrow with its teeny, tiny head? It'll tear off a teeny, tiny specimen of your tissue and then float in the water till I retrieve it and take it to the lab for genetic and hormonal analysis. Namely, knowledge. Shared. From a tiny bit of DNA. So what do you say?

 SHE *lowers the crossbow.*

Not yet?

Okay. Then let's try the old-fashioned way, a little convo. I'll bring the Merlot, you bring the . . . protozoa. Do you actually enjoy eating plankton? I mean, what's the attraction, besides the fact that it keeps you alive? It can't taste good, and it has no umami.

Though who am I to argue? I'm glad you don't eat women for breakfast. You don't, right? I mean, every time a whale swallows a human, he spits 'em out.

Right?

I mean . . . okay a lot of that is hearsay.

Though that lobster diver last year—his friend witnessed the whole thing. And Jonah—well, you just have to trust God. I guess.

I know officially you're an omnivore, but minnows don't count as meat. So you're practically a vegetarian. I respect that. It's environmentally responsible. But it does make me wonder: Where do you find joy? No booze, no blood . . .

Yeah, sure, exercise. Well, if I could spyhop, I'd exercise, too.

    SHE *pantomimes a whale leaping out of the water.*

All those endor*phins*, pun intended. When you leap from the water and arc in the air, you look like you're flying, flapping your fins like wings. You look ecstatic. And it's infectious. I feel it, too. Though I wish you'd hang out a sec so I could shoot you.

It isn't gonna hurt.

Okay maybe a tiny bit, but think about it: All I need's a few cells, a tiny bit of flesh. Not even an ounce. You weigh 128 million ounces. Surely you can spare one.

I'm a tiny human, so an ounce would be a bigger deal to me proportionally, but if I had to give up one ounce to save the world, I'd do it. Even if it hurt. For the good of mankind.

Okay. I hear you. Who cares about mankind? For the good of whalekind. Earthkind! Seriously. We are talking about the future. Our legacy. Yours and mine.

I know. I know.

What's the point of legacy if the world's gonna end?

Who cares if you're in the history books if there's no one left to read them?

But it's still worth a try. Don't you think?

These aren't my official questions, by the way. This is small talk. Just to, you know, ease you into a more meaningful exchange. Since you don't drink Merlot.

My real question, the reason I'm here, what I need to know is . . .

Why don't you have kids?

I realize it's a sticky subject. Personal. Sensitive. Sometimes when you ask people why they don't have kids, they start to cry. Which is why I try to restrict my conversations to sea creatures.

And no one should have to justify *not* having kids. Not with all the overpopulation, starvation, lack of resources. And you know, unwanted children are more likely to become criminals! Is that what you're afraid of, that your kids might be drug dealers or mass shooters? Not that you're afraid of anything at your size. But you should be. You're in the greatest danger of all.

If it's your choice, then I respect it. I get it. You're enjoying your personal freedom, exploring your identity. You find fulfillment in other ways; you have passions and plans. And kids are so needy. And clingy. And loud. And maybe you're not attracted to males. Or maybe he just didn't sing the right song.

This isn't hypocrisy or dumb curiosity. I'm a scientist. So if you'd prefer for me to phrase my query in a more scientific fashion, I can do that:

As a female *Megaptera novaeangliae* of procreative age, why have you failed to produce a calf?

No, not *failed*. I don't mean to find fault. Except . . . typically we spot between 12 and 18 calves this time of year. This year, we've spotted only four.

That's at least a 67-percent reduction in the birth rate. In one year. That's a terrifying trend.

Whales reduce greenhouse gas levels in the atmosphere.

Whales maintain the balance of the ecosystem.

Whales provide the air that we breathe.

We need you. The world needs you. And by *you* I don't mean just Penelope but your progeny. You have to have kids. Or at the very least, tell me why you don't.

Because while we don't need more humans, we do need more whales.

Odessa *aims the crossbow at the ocean.*

Please let me shoot you.

Tableau.

PART TWO

Challenging Traditions through the Science Performance

4

Spooky('s) Action at a Distance

Remixing Science and Performance in the Planetarium Show

Mike Vanden Heuvel

The philosopher and historian of science Michel Serres described the relations between the hard sciences and the human sciences by noting that

> The best light is obtained in the mingled region of interferences between two sources, and this region vanishes if the two flows have no common intersection. If each center claims to be the sole source of light . . . then the only compass readings or pathways obtained are those of obedience.[1]

Such mingled regions provide ample opportunities for theatre and science to productively interfere with one another, yet the history of such collaborations shows they are seldom explored. Perhaps owing to a long-standing desire to bridge the chasm between the "two cultures" (to use C. P. Snow's timeworn conceit), interdisciplinary collaborations between science and theatre have proceeded more cautiously than other art forms, with theatre often assuming the position of obedience to the epistemic authority of science.[2]

A paradigmatic expression of such relations is the conventional planetarium show, which traditionally assembles an audience of science-curious spectators to passively consume an explicitly theatricalized form of science. Both by staging the planetarium lecturer or voice-over as a charismatic source of scientific authority and also endowing them with Prospero-like powers to call forth the immersive wonders of the universe in stunning visual projections, such shows maintain whiteness as the source of imperial power to scientifically map and colonize the visible universe. Performance, meanwhile, is typically positioned to render astronomy spectacular while muting its own powers to engage critically with such modes of representation.

Such demonstrations have certainly evolved since the purely didactic shows produced in the hundreds of grammar and high school planetaria constructed after Sputnik launched in 1957.[3] Attempts have been made to enliven the science-art intersection involved in planetarium offerings and to introduce more nuanced theatrical elements to destabilize the one-way relation of authority. In what follows, I look briefly at three recent attempts to restage such shows, all of which seek to decenter science as a fixed discourse being presented uncritically to the typically passive planetarium audience: I first examine Sound&Fury's production of Hattie Naylor's UK stage play *Going Dark* and Nina Wise's US multimedia planetarium show *The Kepler Story* as related examples of works that attempt to deconstruct the discourse of the planetarium show but ultimately fail to evade the position of obedience—fail to make performance, as it were, disobedient in its relation to science—and thereby open more dynamic pathways for science-theatre collaborations.

But by making room in the white sphere of the planetarium for what George Lewis terms "Afrological" traditions, the Black hip-hop turntablist, dubstep composer, and multimedia artist Paul D. Miller (whose "constructed persona" is DJ Spooky, aka that Subliminal Kid) productively interferes with entrenched channels of communication between science and its public.[4] Rather than follow the traditional pathway of science-art collaboration, Miller resituates his work at a crossroads, a strategy derived from blues traditions, to trivialize (L. adj. *trivialis*, "of or belonging to the crossroads") and remix relations between science and performance and thus render these truly interdisciplinary.[5] His collaboration with the Neukom Institute of Computational Science at Dartmouth College led to the creation of *The Hidden Code* (2016). The title might conjure up the image of a Dan Brown novel or a work of conventional science drama focused on genetics or, like Hugh Whitemore's Alan Turing play *Breaking the Code* (1986), on computation and the lifestyle "code" by which Turing sought to avoid discovery as a gay man. In all such cases, the prospect would entail uncovering what the code conceals, decrypting its mysteries through ratiocination and bringing it into clearer view for historical or ethical judgment. But in Miller's work, the hidden code designates not displacements in signification to be untangled but rather acts that produce what he calls "rhythm science," defined as practices that undo stable denotation by virtue of being transformed into code, the binary state of digital information that allows it to be hacked, sampled, remixed, and recontextualized within new contexts, or rhythms, of meaning.[6] Remix has evolved as a potent method for intervening in colonial and racist performances, as Miller showed in his best-known multimedia remix, *Rebirth of a Nation* (2007), which deconstructed the notorious D. W. Griffith 1915 classic *Birth of a Nation* while also channeling the title of the 2006 album by Public Enemy.[7] Like much remix art, the central strategy involves treating the source film as a found object and thus undermining the intellectual property of the original, thereby redirecting its meanings. In the case of *Rebirth of a Nation*, this included sampling blues music and other Black art forms over the film to restore the sounds of history that Griffith's film suppressed. In *The Hidden Code*, DJ Spooky similarly transforms the planetarium show into an event that eschews building bridges between art and science in favor of bringing them into a relation of productive interference.

The strategy is disruptive because all too often attempts to bridge the two cultures between science and theatre end up serving what Andrew Barry and Georgina Born define as interdisciplinary "logics" ("different kinds of rationale, motivation, or justification for interdisciplinary practices") that subdue rather than energize collaboration.[8] In their review of art-science practices in the UK (which largely parallel those in the US and other Western neoliberal societies), Barry and Born show how efforts to bridge the two cultures have typically produced logics of "accountability" and/or "innovation."⁹ These primarily, if not exclusively, served the needs of science. The logic of accountability concerns the "ways in which scientific research is increasingly required to make itself accountable to society" and so typically deploys the arts to assemble a public for this apologia.¹⁰ For the logic of innovation—the pressing need for science to display its steady contributions to technoscientific progress and the neoliberal knowledge economy—the arts provide aesthetic elaboration and affective triggers to stimulate that display.

Given these prevailing logics, it's only natural in neoliberal societies to find NGOs, foundations, and corporations flocking to support endeavors using the arts to abet science, often for the public communication of science but occasionally for commendable interdisciplinary attempts to solve complex problems related to, for instance, global health care or climate change. The advent of "sciart" as a field, then, develops symmetrically with the rise of organizations in the United Kingdom and the United States like the Wellcome Trust, the Calouste Gulbenkian Foundation, the Alfred P. Sloan Foundation, the National Endowment for Science, Technology, and the Arts, and the like. Once the funding becomes centralized in such science-based bodies devoted to the public understanding of science, Barry and Born insist, support flows to projects that emphasize making science more accountable *to* the public and more innovative *for* the public:

> In these developments, art that is in dialogue with science is understood as a means by which the (absent) public for science can be assembled or interpellated. Science is conceived as finished or complete, and as needing only to be communicated, understood, or applied, while art provides the means through which the public is mobilized or stimulated on behalf of science.¹¹

The rise of science communication breathed new life into the imperative to bridge the arts and sciences, and theatre has played a significant role in this endeavor. However, theatre's position as a medium to communicate science, but not one with the potential to interfere with or engage critically with science, was usually reified. As Born and Barry conclude:

> But—and a final point—whether motivated by accountability or innovation, in the trajectories of UK artscience that we have

sketched, interdisciplinarity is uniformly conceived in the terms of what we have called the service-subordination mode, auguring hierarchical relations in which art is enrolled in the service of science.[12]

Ultimately, my goal in this chapter is to locate practices that go beyond these limited logics and modes of collaboration and explore how the staging of these select planetarium shows pursues (with different degrees of success) what Barry and Born call a "logic of ontology." This is described as "an orientation in interdisciplinary practices towards effecting ontological transformation in both the object(s) of research and relations between the subjects and objects of research."[13] To explore such transformational dynamics, I first return to the work of Michel Serres, paying particular attention to his metaphors of the Northwest Passage and the parasite to analyze how he imagines a particular kind of ontological space for transforming flows between the hard sciences and the arts.

First, it is worth pausing here to ask whether the metaphor of the bridge, itself, might bear responsibility for restricting how we continue to think through relations between theatre and science. The bridge metaphor, as Barry and Born (among others) argue convincingly, continues to exert power today, sometimes via the use of more contemporary terms like "interface," which is similarly suspect for conveying that the junction between interdisciplinary domains must be perfectly engineered and under control to be productive. Bridges, after all, are stable structures and the relations they make possible are fixed and linear, reversible in the way Newtonian equations and mechanical time function: it hardly matters in what direction in space and time the traffic flows over a bridge, the relation moves either way—but, importantly, *only* either way. The itineraries it makes possible are thus chartered, ordered, and efficient to the extent that few accidental or random effects are left to possibility. And even apart from the way that bridges enforce restricted forms of passage, the real problem with such structures is that they lift us over the channels through which wild currents flow: according to Serres, those turbulent flows, noisy as they are, are worth attending to.

For Serres, the space between the knowledge domains of art and science is not geometric and stable but complex, likened to the shifting currents of the labyrinthine Northwest Passage that links the Pacific and Atlantic oceans: "with shores, islands and fractal ice floes. Between the hard sciences and the human sciences, the passage resembles a jagged shore, sprinkled with ice, and variable . . . It's more fractal than simple. Less a juncture under control than an adventure to be had."[14] Space, both geographical and discursive, is for Serres topological, which Steven Connor describes as "geometry plus time, geometry given body by motion."[15] Connor argues that Serres's understanding of topology shifts the view from fixed or Euclidean "form" to a focus on shapes that undergo trans*form*ation as regular figures are squeezed, folded, stretched, and kneaded to produce more complex contours while still maintaining their fundamental structure. Not concerned, like Euclidean geometry, with measurement, topology is the study of relations, disjunctions, and connections that shift over time. The Northwest Passage, similarly shifting and mensuration-defying, becomes for Serres a metaphor for interdisciplinarity that has no place for bridges: as Connor has it, "[i]nstead of carefully mapping the terrain, Serres traces out unpredictable, even vagrant itineraries through landscapes not given in advance" and so confounds any simple choice of a preconstructed route.[16] And without the possibility of a fixed map, there's no basis for calling in the engineers to bridge the channels.

The complex navigation Serres asks us to endure traversing the Northwest Passage forces us to become more robust in our interdisciplinary explorations. For him, disciplinary and conceptual divisions can only be encountered productively by exploring the channels between them, passages through which energy and information run constantly but unsteadily and entropically, with the ever-present risk of loss of order and coherence. That Serres grounds his thought in thermodynamic concepts is hardly surprising, given his interest in identifying what he calls "thermal exciters" in physical systems and discursive relations.[17] These operators always transform the medium of exchange of which they are a part, acting as surges of turbulence or energy that stir up a system in equilibrium by raising the temperature or making unexpected movements or noise. The threat to stability within such increasingly far-from-equilibrium systems couldn't be clearer, yet Serres formulates a means by which such distortions might produce invention and forge new kinds of relations.

In Serres's thinking, the thermal exciter not only irritates and inflames but also makes its presence known by interrupting channels of communication with clamor and noise. Another key Serrean concept, noise is what entices us to descend from the stable bridge into the turbulent flows of the channels below. Now, however, the channel transforms into a message relay, a communication channel. For Serres establishes a crucial third state of time and space beyond *form* and trans*form*ation, that is, the space of *inform*ation where messages, data, codes, and conceptual categories undergo similar forms of mixing.[18] Form; transform; inform: the first is associated with Newtonian mechanics and the stable space of Euclidean geometry, as well as reversible time via the first (conservation) law of thermodynamics; the second is the domain of topological space and the second law governing energy transformations, which create the irreversible arrow of entropic time; and the last emerges in the complex flows of information, governed by a form of entropy ("the entropy of the message") but, significantly, not hostage to it. For in information lies the possibility of negentropy, in which muddled messages and entropic time itself become sporadically reversible, allowing new and complex patterns of order to emerge.[19] As we will see, these are commensurate with the strategies of remix as well.

One instance of communication interference provided by Serres is the parasite, who in social exchanges conventionally takes in another's energy (food, sustenance, status, resources) without returning an equal share of work, contributing instead only flattery, lethargy, and hot air. Throughout *The Parasite*, Serres famously reconfigures this thermal exciter into a component of communication exchanges by linking it, via the French *parasite*—which translates to "noise" or "static"—and then reading interdisciplinary relations through the lens of information theory. Here, "noise" is analogous to friction in thermodynamic exchanges, the static contained in every communication interchange that dissipates, or parasites, clear messages. But the parasite, as Serres recasts it, is also the source of a new relation between domains: instead of establishing equilibrium by enabling a stable exchange of messages between sender and receiver, by bridging that channel and making communication clear and predictable, the parasite interrupts that symmetry and produces nonlinear and unpredictable effects. Like noise in an information channel, the parasite muddles the logic of exchange and both betrays the original intention or message and creates the potential for more complex messages and relations by adding its noise to the sum of the information. And Serres maintains that the results of such noisome interruptions can be profound if we think about static or noise as another *kind* of information, rather than as information's chaotic Other: as a source of equivocations that destabilize both parties of an exchange to produce unanticipated effects. Like Barry and Born's ontological logic of interdisciplinarity and like the improvisational methods of jazz and remix, the parasite is, as in Serres's earlier formulation, "less a juncture under control than an adventure to be had."

But not an adventure pursued by the conventional planetarium show. Art hosted in such sites invariably bears the obligation to render science spectacular without interfering with its knowledge. Science takes the shape of knowledge that is finished and complete, a stable discourse only needing to be communicated clearly and entertainingly to be understood. Equally striking, however, is how the conventional planetarium show creates a public *for* science, in the sense that Born and Barry suggest: "art-science promises to assemble a public for science in a form to which science can then consider itself accountable; science is rendered accountable to a public that is, in turn, properly disposed toward it."[20] Theatre's role in this construction of a public sympathetic toward science is to make itself transparent, the relay channel through which the public is mobilized and stimulated to pursue knowledge of, or empathy toward, science.

Given the historical one-sided nature of the relation, it is unsurprising that as science-theatre collaborations have evolved projects situated in planetaria or mimicking their presentations have sought to rebalance the relation. But this hardly guarantees that such interdisciplinary projects will deconstruct the bridge model or produce the robust kinds of interaction that Serres proposes. While more contemporary work may undercut the conventional role that performance is typically assigned of embodying science as a master discourse and rendering its methods and achievements in artful design, they also mostly maintain the goal of bridging theatre and science by adhering to the logics of accountability and/ or innovation.

The UK company Sound&Fury, for instance, produced Hattie Naylor's *Going Dark*, which presents the story of an astronomer and planetarium lecturer, Max, who is losing his sight to Retinitis pigmentosa. Sound&Fury is notable for its use of total darkness and surround-sound design to explore sensory deprivation and its impact on perception, particularly the way the mind will create images even absent optical input. Such effects are linked thematically to Max's struggle to maintain his memories, identity, and relationship with his son as he begins to lose his sight. These resonate effectively with the play's embedded planetarium lectures, during which digital projections and lighting mimic a full-dome immersive experience. Max's audience talkbacks celebrate the progress and innovations that have brought astrophysics to its present advanced state of knowledge concerning the universe. As his disease progresses, however, Max's lectures describe how the universe, itself, is "going blind" as it expands: "No fixed point in a universe that is getting bigger and bigger, rushing away from itself so fast that one day its light will never reach us. The universe itself is going blind—going dark."[21] In several scenes when Max is alone, the audience is left in utter darkness as surround-sound effects conjure the experience of Max's sightlessness on busy city streets. In both form and content, then, the performance denies the powerful scientific gaze that the planetarium show traditionally embodies and addresses poignantly the limits of both individual human sight and the range of scientific knowledge more generally. "The more we find out about the cosmos," Max informs his planetarium audience, "the more we realize what we don't see, what we can't see, and what we will never see."[22] The production thus effectively repositions true vision in the imagination of the spectator rather than solely in the methods and technologies of astronomy.

However, as Alex Mermikides astutely observes, the blackout and surround-sound effects in *Going Dark* are reserved for the representation of Max's visual pathology rather than the scientific content of the play. As she notes, "scientific ideas are explicitly communicated . . . through an essentially classical dramatic format," that is, as an authoritative science lecture.[23] While Max's lectures effectively elucidate astrophysics and the psychophysics of perception, they maintain the conventional planetarium show's emphasis on awing the audience through narrative and spectacle. Science is never pathologized, is never (as Serres would say) interfered with by its being featured in the blackout scenes or disoriented by the surround-sound design. The play thus never parasites the science it presents, and scientific clarity and far-sightedness act, rather, as the ableist counterweight to Max's loss of vision, suggesting that even as the individual may lose sight there are powerful forms of scientific vision and knowledge that will endure. So, while the play presents an interesting intervention in the pedagogical imperative of the traditional planetarium spectacle, it still treats science as an established discourse of knowledge that remains unwelcoming to the parasite.

Even when performance returns to the site of the planetarium proper, we see how attempting to mount a counter-discourse to its historical role can become entangled in the logic of accountability, which dampens the effects of noise and the parasite. Nina Wise's *The Kepler Story* is a multimedia planetarium-sited show based on the biography and ideas, scientific and mystical, of its namesake.[24] The show elaborates science by combining full-dome immersive imagery with live acting, recorded music, high-definition computer animation, and sound design. Marketing for the work emphasized its goals, which include "visualiz[ing] the workings of [Johannes] Kepler's mind so that the audience can actually see Kepler's discoveries as he might have seen them in his own interior reality."[25] The premise is that the view of science in Kepler's own mind may differ from what conventional histories of science have narrated. This perspective thus allows for a critique of science, distancing the show from the goals of the conventional planetarium spectacle. As Wise has written, *The Kepler Story* shows "how the Scientific Revolution removed the understanding of the interconnected nature of reality from our lives and defined the way we conduct our business, practice medicine, consume resources, and create international political policies."[26] Kepler's paeans to the intrinsic harmony of nature offer an alternative to this Baconian expression of instrumentalized science, often by resonating with ideas from contemporary systems theory (the original script was cowritten by Ralph Abraham, an important figure in the study of dynamical systems). By virtue of the powerful immersive experience, the show seeks to endow the spectator with the desire and means to find solutions to "the economic, political and environmental issues of our own day."[27]

However, despite destabilizing the site of scientific display and the planetarium's science-affirming pedagogy, Wise's show still adheres to the interdisciplinary logic of accountability that uses theatrical elements to render science answerable to its public. Kepler's science is reenvisioned (mostly by selectively retrieving material from his mystical writings) and repositioned as a *pharmakon*: a poisoned pill coursing through the veins of the Scientific Revolution that, when it stands revealed in the retelling of *The Kepler Story*, can remedy the toxic dominance that science has established over nature. Science is both the origin of this disease and disenchantment, but also its cure, and the spectacular show testifies that science has from its very foundations supported a systems approach to the environmental and cognitive crises that it, in part, has helped to create.

Thus, despite laudable attempts to rearticulate the use of performance in the planetarium show, the logics of accountability and innovation, and to some degree the service-subordination model of collaboration, have remained in place. But what happens when the relevant agents in a collaborative effort to restage the planetarium show choose to listen attentively to the noise between the channels that separate them and to meet in this turbulent space? When, rather than using spectacle and other theatrical attributes to communicate science as an established *dispositif* to be presented to a public deficient in knowledge or empathy toward science, they instead allow performance to parasite that message?

DJ Spooky aka that Subliminal Kid is the avatar whom artist Paul D. Miller considers a form of "social sculpture," a "persona in the form of shareware," allowing him greater mobility across creative fields. Thus unencumbered, he moves freely across music (electro, hip-hop, illbient, jazz), conceptual art, installation art, digital and video art, theoretical writing, and other expressive forms.[28] He has established himself as a prolific figure at the crossroads of art and science, having served a number of residencies at science institutes and produced work on climate change and hypsography.[29] As mentioned, the concept of the crossroad, based in blues music and mythology, is part of the Afrological sensibility Miller brought to the Neukom Institute for Computational Science at Dartmouth, where he devised and presented *The Hidden Code*.[30] The crossroad, unlike the bridge, is a transient, liminal, multidirectional space of unpredictable movement: "Less a juncture under control than an adventure to be had."

The Neukom residency afforded the rare opportunity for a robustly interdisciplinary science institute devoted to studying the impact of computation across many technical and social fields to mingle with a performing artist whose working methods, heavily invested in digital media and aesthetics, aligned with its mission.[31] Such a complex interaction can only take place where one domain in an interdisciplinary exchange cannot dominate the other but rather where both are mutually parasited. Given the computational focus of the Neukom Institute and the performative practices of remix deployed by DJ Spooky, we might think of the collaboration as open source. Indeed, commenting on Miller's 2004 book *Rhythm Science*, Lawrence Lessig notes this analogy with the universe of infinite file sharing and remixing that characterizes Miller's aesthetic:

> We've ended the century of broadcast culture—when manufacturers produced the culture we consume. In this brilliant and beautiful book, Paul Miller gives us the rhythm of sampled culture—culture created by those who can remix, and by technologies that enable anyone to remix. *Rhythm Science* is science; it is art; it is the story of how freedom would build better science and art.[32]

Miller traces his practice as a remix artist through a variety of traditional Black, African, and Caribbean expressive traditions: blues music, the griot tradition of storytelling through music, jazz performance, and others.[33] As well as providing methods for rejecting or intervening in Anglo-European aesthetics and discourses, these all repurpose found material and trouble the notion of an original source of inspiration, thus presaging the open-source culture of remix. As Jesse Stewart writes, "the multiplicities and flux associated with blues geographies now play out on a global scale in the art of digital file transfers and a virtually limitless digital archive."[34] In the world of Web 2.0 where DJ Spooky operates, everything has become code that can be easily transferred, archived, reconstituted, recontextualized, and reperformed.

Sometimes, however, with proprietary systems and information—perhaps even those that make up the formidable edifice of science—the code must first be hacked. For Miller, the archive for *The Hidden Code* included his research into the covert codes created by video game "crackers" (such as the "vectrex vectorscope" hacks that allow gamers to repurpose old Atari consoles) and Stephen Wolfram's Alpha Generative Engine, used to repurpose artistic techniques into algorithms to be used in DIY computer games and SFX systems.[35] Miller seems to have sensed a correlation between the creativity involved in hacking and the Neukom Institute's own network of disciplines—neuroscience, quantum computation, emergent systems theory—because, like hacking and remix, all these sciences are engaged with the transformation of fixed or binary structures into more complex systems. Miller thus treated the sciences of the Institute equally as material to be hacked and interfered with, rather than stable scientific knowledge. Rather than passively maintaining the traditional epistemic authority of science, he treated it like any other form of code, as something to be sampled and remixed. In this role, DJ Spooky is positioned as Serres's thermal exciter. Instead of assuming the conventional role of the guest of the Institute and taking the part of an equal interlocutor in an art-science exchange predicated on bridging the domains, Miller adopts the position of the parasite, the source of disequilibrium. And like any good house artist, the DJ brought the noise.

The performance of *The Hidden Code* took the form of an immersive, multimedia show featuring projection technology (with visuals created by Boston's Hayden Planetarium) along with live and sampled music and narrated performance, thus similar in many ways to the staged planetarium lectures of *Going Dark* and the full-dome experience of *The Kepler Story*. But significant differences point to decidedly different goals. In *The Hidden Code*, the science collaborators involved in the show do not take center stage or speak in the voice of authoritative experts. Two of the performers are, in fact, well-known scientists: the physicist Stephon Alexander (author of *The Jazz of Physics* (2016) and a professional jazz musician) plays saxophone and speaks about relations between "jazz thinking" and high-energy physics. Marcelo Gleiser, a physicist and cosmologist (as well as another science popularizer), reads from his poetry and books such as *The Tear at the Edge of Creation* (2013) and *The Island of Knowledge* (2014).[36] These conventionally authoritative figures are, first, appearing as much as performers as they are scientists, which mediates their position as expert science consultants in interesting ways (such as staging them within the poetics of failure that attends all live performance).

But here the actors using discursive language are not the primary focus: the DJ runs the show, and his computerized turntable activates and alters the network in real time. The scientific content of the physicists' speech is mediated and redirected by the visuals and immersive environment, which interrupt and recontextualize them. There is no determinable relationship between the visuals and the recitations, as each was created separately and then remixed in the performance: thus each performance would result in a different remix. Most importantly, the scientists' performances become themselves part of the codes being remixed by DJ Spooky. He uses real-time digital editing software to actively sample and alter their voices and Alexander's instrument from the DJ turntable, mixing these with music samples and found acoustic material from the digitized archive available on his laptop. Here, we reach our final "channel," now describing the multichannel audio platforms used by DJs to play "stems" of digitized beats and loops to isolate different groupings (instruments, vocals, timbres) and then to remix these to create a new track utilizing the DJ's traditional interventions: breaks, needle drops, looping, scratching, backspinning. This follows, then, not the linear, representational dramaturgy of *Going Dark* or *The Kepler Story* but rather the redirective practices of remix, which introduce appropriated samples only to juxtapose them in new alignments to create varying cognitive and embodied beats, loops, and rhythms. The goal is not scientific concepts to be recalled and repeated as fixed knowledge but to create new negentropic flows: the kicks and snares that emerge when the spectator finds a convergence of sound, voice and visuals they find meaningful. "Rhythm science."

The parasite/DJ exchanges the orderly information of the planetarium lecture for disorder, pumping out into the surroundings not systematic representations of facts and stable scientific knowledge but a kind of entropy—difference, disorder, unpredictability—interrupting and redirecting the scientific content while never overwriting it completely. This is quite unlike the sound design of a conventional planetarium show, which aligns uplifting music with inspiring visuals to move the audience in consonance with the scientific content.

The Hidden Code thus moves decidedly in the direction of Barry and Born's ontological logic, which effects "transformation in both the object(s) of research and relations between the subjects and objects of research."[37] DJ Spooky's sensibility as a remix artist aligns with the interdisciplinary formation of the Neukom Institute insofar as both translate or transform the space of the planetarium from a location of stable scientific knowledge waiting to be communicated to the public to a performative site of what Barry and Born term a "public experiment."[38] They identify recent shifts in sciart collaborations in which practitioners drawn to such interdisciplinary projects emerge from backgrounds in "conceptual and post-conceptual art, including performance, installation, public and activist art."[39] (Miller has been active in all these fields, having exhibited at the Whitney and Venice Biennales and regularly referencing in his work and writing the work of Duchamp, Cage, Beuys, and others.) These have, in Barry and Born's view, altered the direction of some science-performance collaborations away from the strictly subordination-service model to "an orientation to what we have termed the logic of ontology, and in some instances a concern with the production of what we will call public experiments."[40] Like other forms of relational art, *The Hidden Code* transforms the space it occupies by activating its potential as a site of social relations, rather than knowledge transfer, thereby experimenting with the public communication of science. Here, spectators are empowered to respond to the performance parasitically, in the sense they are invited to enter the turbulent channels between art and science to navigate its unpredictable flows of information, not to find the path of an obedient understanding of physical facts but to feed on the noise and to seek the myriad patterns of meanings and pleasures it contains.

This is significant because it hails a new plural form of publicness, one that situates the spectator in a different relation to its knowledge of science. Rather than the teleology of transmitting a putatively stable formation of scientific knowledge to a public lacking it, remix operates by another logic, one that sanctions the public's active discovery of new relationships and unforeseen constellations of meaning. Spectators establish connections and patterns from the remixed texts, sounds, and images in a manner not unlike the organized nonequilibrium states that Serres invokes, patterns rich in potentialities yet maintaining their own complex fractal structure. Rather than narrowly framing scientific ideas so as to be a window for the spectator's understanding, remix multiplies the frames (like a Windows desktop) to allow multiple cognitive trajectories and patterns of recognition to emerge. While not beyond critique, for instance in the way *The Hidden Code* maintains an essentially frontal, proscenium perspective that limits audience engagement with the public experiment, the work produces a multidirectional and turbulent channel—rather than a bridge—into which spectators may immerse themselves.

Notes

1. Michel Serres with Bruno Latour, *Conversations on Science, Culture, and Time* (Ann Arbor: University of Michigan Press, 1995), 176.
2. C. P. Snow, *The Two Cultures and the Scientific Revolution* (Cambridge: Cambridge University Press, 1959).
3. See Jordan Marche, *Theaters of Space and Time: American Planetaria, 1930-1970* (New Brunswick, NJ: Rutgers University Press), 2005.
4. George Lewis, "Improvised Music after 1950: Afrological and Eurological Perspectives," *Black Music Research Journal* 16, no. 1 (2002): 91–122.
5. *Online Etymology Dictionary*, etymonline.com/word/trivial.
6. Paul Miller, aka DJ Spooky that Subliminal Kid, *Rhythm Science* (Cambridge, MA: MIT Press, 2004).
7. For a discussion of the methods employed in *Rebirth of a Nation*, see for instance Jason C. Apple, "Re-mixed Histories: Cinematic Narrative, Found Footage, and Hip-Hop Historiography in DJ Spooky's 'Rebirth of a Nation,'" *Spectator-The University of Southern California Journal of Film and Television* 26, no. 2 (2006): 45–53.
8. Andrew Barry and Georgina Born, "Art-Science: From Public Understanding to Public Experiment," in *Interdisciplinarity: Reconfigurations of the Social and Natural Sciences*, ed. Andrew Barry and Georgiana Born (London: Routledge, 2013), 247–72.
9. Andrew Barry and Georgina Born, *Interdisciplinarity: Reconfigurations of the Social and Natural Sciences* (New York: Routledge, 2013), 14.
10. Barry and Born, "Art-Science," 249.
11. Ibid.
12. Ibid., 255.
13. Ibid., 249.
14. Serres, *Conversations on Science*, 70.
15. Steven Connor, "Topologies: Michel Serres and the Shapes of Thought," *Anglistik* 15, no. 1 (2004): 106.
16. Ibid.
17. Michel Serres, *The Parasite*, trans. L. R. Schehr (Minneapolis: University of Minnesota Press, 2007), 191.
18. Michel Serres, *Atlas* (Paris: Editions Julliard, 1994), 83.

19 For definitions of terms like "negentropy" and a detailed account of information theory's evolution, see N. Katherine Hayles, *Chaos Bound: Orderly Disorder in Contemporary Literature and Science* (Ithaca, NY: Cornell University Press, 1990). See especially "Self-reflexive Metaphors in Maxwell's Demon and Shannon's Choice: Finding the Passage," 31–60.

20 Born and Barry, "Art-Science," 254.

21 Hattie Naylor in collaboration with Sound&Fury, *Going Dark* (London: Methuen Drama, 2012), 32.

22 Ibid.

23 Alex Mermikides, "Taking Direction: New Dramaturgies of Science from the *Splice Symposium* and the *Performing Science* Conference," *Interdisciplinary Science Reviews* 39, no. 3 (2014): 292.

24 The show premiered at the Morrison Planetarium, part of the California Academy of Sciences, in San Francisco in 2013.

25 Whitney Dail, "Experiencing Science through Immersive Theater: A Q&A on The Kepler Story," National Endowment for the Arts, 22 May 2012.

26 Ibid.

27 Ibid.

28 Paul Miller, *DJ Spooky: Paul D. Miller aka That Subliminal Kid*, https://djspooky.com.

29 Among his numerous interdisciplinary residencies, Miller has produced work with the Norwegian University of Science and Technology (2020–1) and recently served as Artist-In-Residence at the Yale University Quantum Institute (2021–2).

30 Jesse Stewart, "DJ Spooky and the Politics of Afro-Postmodernism," *Black Music Research Journal* 30, no. 2 (Fall 2010): 342.

31 For the Neukom Institute's mission statement, see https://neukom.dartmouth.edu/about/about-neukom-institute/mission-overview.

32 Lawrence Lessig, "About the Book," https://mitpress.mit.edu/sites/default/files/titles/content/mediawork/titles/rhythm/rhythm_book.html.

33 Jesse Stewart, "Review: *Rhythm Science*," *Critical Studies in Improvisation* 1, no. 1 (2004): 1.

34 Ibid., 343.

35 Paul Miller, "World Premiere Performance: The Hidden Code," *DJ Spooky: Paul D. Miller aka That Subliminal Kid*, djspooky.com/the-hidden-code/.

36 Stephon Alexander, *The Jazz of Physics: The Secret Link between Music and the Structure of the Universe* (New York: Basic Books, 2016); Marcelo Gleiser, *A Tear at the Edge of Creation: A Radical New Vision for Life in an Imperfect Universe* (New York: Simon & Schuster, 2010); Gleiser, *The Island of Knowledge: The Limits of Science and the Search for Meaning* (New York: Basic Books, 2014).
37 Ibid., 249.
38 Barry and Born, "Art-Science," 250.
39 Ibid., 255.
40 Ibid., 257.

5

Using Short Digital Films to Counter Stereotypes about Scientists of Color and from Marginalized Backgrounds

Mónica I. Feliú-Mójer

For many people, media—particularly digital media (online newspapers and magazines, videos, social media sites, etc.)—is the primary source of information about science-related topics, including who scientists are (i.e., their backgrounds and identities). Indeed, more and more people in the United States rely on the internet to learn and stay informed about science and technology; according to the Pew Research Center, more than half (55 percent) of US adults report seeing science-related posts on social media.[1] The majority of adults in the United States get their science news from general news outlets,[2] many of them presumably online. A large majority (82 percent) of US adults say they often or sometimes get news from a smartphone, computer, or tablet.[3]

Unfortunately, no matter the platform, media recurrently conveys narrow views of who is a scientist and how people become one.[4] Science, technology, engineering, and math (STEM) characters present in film, TV, and streaming tend to be male (62.9

percent) and overwhelmingly white (71.2 percent).[5] Additionally, scientists are stereotypically portrayed as "mad" (e.g., with crazy hair like Doc from *Back to the Future* and Beakman from *Beakman's World*),[6] nerdy and lacking social skills (e.g., several characters in *The Big Bang Theory*),[7] and wearing lab coats.[8] These limited representations reinforce harmful notions that people who are not male or white are uninterested in science or incapable of becoming scientists.[9] They also misrepresent the diversity of interests, personalities, and styles of people in science. These factors combine to perpetuate the historical and continued underrepresentation and exclusion of people from marginalized backgrounds (e.g., minoritized racial and ethnic groups, people with disabilities, LGBTQ+ and gender nonconforming, and women) in science.

When media platforms share the stories of scientists of color and from marginalized groups, such scientists are frequently presented from a deficit perspective, overcoming great odds, and succeeding in science *in spite* of challenges. We cannot deny the many barriers that make the paths of certain individuals into science harder. However, viewing scientists of color and from minoritized groups through the underdog lens renders their experiences one-dimensional and ignores the systemic oppressions and power dynamics that marginalize them. In her 2019 essay "Deprogramming Deficit," Monica Ridgeway shares how poverty positively impacted her and gave her skills to successfully navigate academia.[10] She recounts being tokenized as a Black student or being told she received a fellowship because she was "a minority." Importantly, she challenges common perceptions about people living in poverty. She shares she had a happy childhood, that science and nature were constants in her life, and that she developed some of the skills that make her an outstanding scholar (e.g., observation, solving problems, making connections) growing up in public housing.

The presentation of counter-stereotypical examples of scientists can shift students' often trite ideas about them (e.g., you must be a genius to be a scientist; scientists don't come from diverse countries and backgrounds) and make scientists more relatable.[11] That's why in 2017, in my role as a producer with Wonder Collaborative,[12] the filmmaking arm of the Science Communication Lab (SCL),[13] I led the creation of a series of short films that we called "Background to Breakthrough" or B2B.[14] The original idea was to produce digital videos featuring scientists of color and from intersectional and historically marginalized backgrounds in biomedical sciences—to show that science careers can be reached via many paths.

The concept of the series was partially influenced by my background. As a woman born and raised in a working-class, rural area in Puerto Rico, I encountered the lack of role models and educational disparities faced by many marginalized individuals. While I was applying to graduate school at Ivy League universities, I was told I would be accepted anywhere because I was a "double minority" (woman and Puerto Rican/Latina). This "advice" presumed I would be accepted into programs because I checked multiple diversity boxes and not because of my merits. Based upon many conversations with friends and colleagues, I also knew that my experiences are not uncommon—and unfortunately, not the worst of what marginalized folks in science deal with.

As I went deeper into the background research for the series, I had an "aha!" process (it went beyond just a moment). I realized that we could explicitly explore the connection between someone's experiences, their decision to become scientists, and their research breakthroughs to challenge common deficit narratives about minoritized scientists. I recognized we could partner with marginalized scientists and experiment with storytelling techniques (animations, documentary-style interviews, etc.) to capture the complexity of their narratives. Most importantly, B2B could catalyze important discussions about diversity, equity, and inclusion (DEI) in science, providing new insights about how filmmaking can change misconceptions about scientists of color and from marginalized backgrounds. Later in the chapter, I examine how we produced B2B videos to counter oversimplified notions of who is and who becomes a scientist and how science is performed, as well as indicators of scientists' impact. By sharing the process of producing and releasing B2B, I demonstrate how the practice of science filmmaking—and more broadly, the practice of science communication—is a valuable form of inquiry that can demonstrate effective ways to share the excitement of science and challenge its status quo and thus establish B2B as an example of Practice as Research (PaR).[15]

Three Stories, Three Different Approaches

B2B is a series of five short videos, roughly four to ten minutes long, about how the culture and backgrounds of scientists of color and from marginalized groups fuel their ingenuity and approaches to problem-solving. The films portray three scientists (Esteban González Burchard, Rebecca Calisi Rodríguez, and Manu Prakash) and how they have succeeded because (and not in spite) of who they are. We chose the three of them for a few reasons. We wanted scientists who represented diverse fields and backgrounds. I was familiar with their work or personally knew them, which facilitated filming. Finally, they were all within driving distance from the San Francisco Bay Area, where SCL is based, which helped keep our costs down.

The films focus sharply on the scientists' identities and life experiences, exploring how these have contributed to each scientist's innovation and excellence. All B2B videos are freely available for streaming on YouTube[16] and the Wonder Collaborative and iBiology websites, where they can also be downloaded. All videos are captioned in English and Spanish (except Manu Prakash's, only available in English).

Esteban González Burchard, MD

Esteban González Burchard is a physician-scientist and professor studying health disparities in asthma among minoritized populations in the United States at the University of California, San Francisco (UCSF). The first film, titled "Who's Included?,"[17] introduces Burchard's research and investigates science's history of racism and exclusion of minoritized populations from clinical and biomedical research, both as study participants and the people leading the research, and how that impacts the health of these groups. Burchard offers the example of a genetic test used to assess whether someone is at risk for an enlarged heart to illustrate how the failure to include participants from diverse backgrounds in clinical research can have devastating consequences. The test was developed mostly in white people and thus failed to account for important genetic variations that are common in Black people. Researchers later discovered that this led to people being misdiagnosed and overtreated for a condition they didn't really have.[18] Burchard shares in the video that, for example, African American athletes were told (unnecessarily) to stop doing any strenuous exercise and to use antiarrhythmic drugs or implanted defibrillators.

In the second video, "Which Box Do I Check?,"[19] Burchard reflects on the role of race in medicine. He recalls a conundrum he faced treating a mixed-race patient when he was a pulmonary medicine resident at UCSF that underscores these issues:

> I saw a patient who was African American, light skin, looked like [Barack] Obama, and he had an occupational injury, where he inhaled toxic fumes. At the end of the test, it came down to comparing him to a reference standard. Depending upon the pull down, I could change the outcome of his results. If I called him Black, he would not have qualified, if I called him white, he would have qualified for benefits. The technician said, well, he looks Black. And I said, well, clinically he's 50-50, so we'd get it wrong either way. And so, I erred on the side of making him qualify for disability benefits. I labeled him as white.

While the details about the patient and the disability benefit laws in this case are beyond the purview of this chapter, with this example Burchard illustrates one of the many ways in which systemic racism manifests in medicine: reference standards used to make important clinical decisions are frequently based on white patients. This exclusion affects not just the health but also livelihoods of people, especially those from minoritized populations.

The third and last video, "An Inclusive Future,"[20] centers on Burchard's background. He is Latino and identifies as Mexican American. He was raised by a single mother in a low-income, predominantly working-class neighborhood in San Francisco. His father was an alcoholic. Esteban's interest in biology came through fishing, which he would do next to the bar where his dad drank. He got into fights and was kicked out of high school. Because of these factors, many would consider Esteban unlikely to succeed in life, let alone in science. But, instead of framing him as an underdog, my team and I sought to draw a direct line between his life experiences and identities and his research. During the interviews, Nona (Nina) Griffin (the film series' co-producer and editor) and I encouraged him to share personal anecdotes that would demonstrate how his unique perspectives inform his work. He shares one from his time as a fellow at Harvard University:

> When I was at Harvard, these older white men had been scratching their heads, asking about this prevalence and severity of asthma amongst Latinos or Hispanics in the United States. They showed that asthma prevalence and death rates were three times higher for Latinos living in the Northeast compared to Latinos living in the Midwest, the South, and the West. And, immediately, I said, "Ha. That's Puerto Ricans versus Mexicans!"... I said, "You know, that is African ancestry coming through the Puerto Rican population increasing the risk of disease."

Burchard also documents the discriminations he has faced, like when a medical school professor told him he didn't have the "right cultural background" to be accepted into residency at Harvard University or UCSF (he trained at both institutions). He shares how important mentorship has been to overcome challenges and succeed in a science career. Burchard talks about how after his parents divorced, his Chinese neighbors took him in and became like family. He proudly mentions how he was strongly influenced by his college wrestling coach, an African American elite athlete getting a PhD, and his Jewish housemates in medical school. He ends this video by reflecting on how his multicultural life experiences influence his research interests and mentoring and how his work is an affirmation of who he is.

Rebecca Calisi Rodríguez, MS, PhD

Rebecca Calisi Rodríguez, a neurobiologist, artist, activist, and associate professor at University of California Davis, is profiled in a video called "Charting an Original Path."[21] The film begins with her introducing her intersectional identities: a Mexican American-Italian woman and mother who grew up in a ranch at the Texas-Mexico border where her interest in animal behavior began.[22] "Charting an Original Path" traces her journey from artist to biologist. While she was painting a commissioned mural at the Dallas Zoo, Calisi Rodríguez was recruited to help understand why the okapi, a mammal related to the giraffe that lives in the Democratic Republic of Congo, was not breeding. She recalls:

> [T]hey said, every time it urinates, collect that urine. So, that was my very glamorous job, but what ended up happening is they found out that their cortisol levels were very high and they were high all the time . . . This might be why they weren't breeding. So, what they did is, they found all the things that could possibly be stressing out the okapi. Remove them, and wouldn't you know? I think it was maybe a month or two later, the okapi started breeding again and to me, this was a huge . . . because, in a way, the okapi were talking to us. They were telling us through their hormones that they were really stressed out so we could do something about it.

The Dallas Zoo project cemented her interest in the connection between the brain and behavior. It also encouraged her to take a seemingly unexpected turn for an artist: pursue a master's and then a PhD in science.

About halfway through the video, she reveals that her research centers on the common pigeon. Her laboratory studies the bird to unravel the mysteries of parental care behavior, as pigeons nurture their young in a way that is analogous to breastfeeding. Pigeons and doves have crop-milk, a substance full of nutrients that—similar to milk from mammals—help younglings survive, produced by both males and females in special cells in the lower esophagus.[23] Calisi Rodríguez's research group studies which genes are active at certain times of important parental behaviors like when feeding crop or pigeon milk, how stress affects reproduction, and how animals adapt to rapidly changing environments.

Throughout the video, Calisi Rodríguez evidences how her scientific and personal identities are closely entwined: "My science influenced me to think about what was happening to my body as a parent . . . [T]his has influenced me to think about what we're researching in science and what questions we're trying to answer." She shares that the time during which early career researchers establish their independent research programs often coincides with their childbearing and caregiving years, a special burden for academic scientists seeking to start a family. Furthermore, academia commonly falls short of supporting parents, as when an important conference in her field (held by the Society for Neuroscience or SfN) provided inadequate facilities for breastfeeding that consisted of "three pop-up curtains and with one really hard, simple conference chair." Rodríguez and others called the conference out on Twitter,[24] tweeting, "This is NOT ACCEPTABLE. We demand REAL change, SFN. U talk the talk, now walk the walk! #WomenInStem."[25] Their denouncements prompted conference organizers to upgrade the breastfeeding area.[26] Moreover, the incident led Calisi Rodríguez and a group of mothers in science to publish an article that offered specific recommendations for conferences to better support parents.[27]

Calisi Rodríguez concludes by reflecting on the hardships of being treated as an outsider and the isolation she experienced during graduate school. She argues for the importance of finding a supportive community to help manage some of the challenges facing marginalized scientists, to mitigate feelings of isolation, to promote a sense of belonging, and to access resources.

Manu Prakash, PhD

The final video of the series, "Finding Sublime in the Mundane,"[28] profiles Stanford University professor of bioengineering Manu Prakash, who is famous for his "frugal science" inventions:[29] low-cost solutions and tools to make science accessible to everyone. Designed for extremely resource-constrained settings, the tools address a range of needs from field diagnostics to hands-on science education. They include Foldscope and Abuzz (described in the video and later in the chapter); Paperfuge, a paper-based centrifuge; and Pez Globo or Pufferfish, an open-source ventilator that can be easily manufactured, created in response to the Covid-19 pandemic ventilator shortage.

Prakash begins by weaving together vignettes of growing up in Rampur, India, in a house with an abandoned chemistry lab in the basement and the time he stole his brother's only pair of glasses to make a rudimentary microscope. A picture emerges: Prakash's life has always involved making things, experimenting, and exploring. He contrasts his creative childhood with the rigidity of high school, where he was required to take entry exams to gain acceptance to good universities. "And these exams were about books," he recalls.

"None of my experience counted, and it was a complete U-turn from my personality."[30] However, Prakash found ways of returning to the open-ended creativity and curiosity of his childhood. Much of his work today is grounded in those early experiences and his fascination with discovering new things in the mundane. Prakash's frugal science efforts are driven partly by the recognition that pursuing seemingly mundane scientific questions is a privilege.

In the film, he shares the origin of two of his famous inventions: Abuzz[31] and Foldscope.[32] Abuzz is a crowdsourced surveillance system that allows anyone with a flip or smartphone to record geolocated sound data about mosquitoes—the deadliest organism on the planet—providing near real-time information about the distribution of different species after events like floods. Viewers experience Prakash's childlike curiosity as he puts together his most famous invention, the origami-like microscope named Foldscope. Foldscope can be manufactured for just a few US dollars, yet it has a magnification of 140x, powerful enough to see bacteria, blood cells, and single-celled organisms.

Prakash concludes the video with powerful advice:

> You don't wait to do your best work, you do it now and you do it with the rigor and capabilities that you bring to the table and you will see it will evolve into something that you really dreamed of. You know, you have to listen to yourself . . . if you don't listen to yourself then you can't really be an honest scientist.

His words underscore a key message of B2B: the value of authenticity and tapping into one's strengths and experiences to succeed in science.

Leveraging Storytelling, Visuals, and Other Filmmaking Techniques

The B2B films are honest and intimate. Each scientist wields their vulnerability to explore moments of wonder and inspiration, of struggle, failure, or insecurity and how they turned them into success. We really wanted viewers to connect with the protagonists, to identity with them, find shared experiences and values, and inspiration in their stories. To accomplish our desired level of empathy and connection, we used a variety of storytelling and filmmaking techniques like speaking directly to viewers, telling anecdotes, providing concrete examples, and using pictures, animations, and secondary footage (commonly called b-roll).

We did in-person, phone, or video preinterviews with Burchard, Calisi Rodríguez, and Prakash. Each of these were at least an hour long and performed by me and the films' co-producer and editor Griffin and executive producer Elliot Kirschner. Our main goals were to get to know the scientists' stories better and get a sense of their personalities and interests. We typically asked them to share their personal and professional journeys and their research. We also discussed filming logistics and asked about previous noteworthy interviews or features that we could review.

We reviewed interviews and other multimedia content and contacted people in our networks who were familiar with the scientists (professionally, personally, or both). One of the challenges of highlighting publicly engaged scholars, who are well known in their fields, is that interviews, articles, videos, and other media are plentiful, making it harder to find new angles to their stories. Together with the preinterviews, the background research helped us understand how our subjects had been portrayed before and which unique perspectives our videos could highlight. For example, Prakash's TED Talk about Foldscope has over 2.3 million views.[33] There are probably thousands of articles, videos, posts, and other multimedia about him and his paper microscope. How could we tell a story that was unique? We discovered that most of what had been said about Prakash and Foldscope focused on the invention itself. We decided that his video would center on his upbringing in India, his culture, his deeply held interests and values, and how they directly inform the frugal science work that he does today.

Preinterviews and background research allowed us to build trust with the protagonists and better understand how we could challenge deficit narratives about scientists of color and from marginalized backgrounds. As I was researching Burchard, for example, I came across a *Buzzfeed* article by Peter Alhous that addressed the former's health disparities research and his lab's efforts to increase the inclusion of people of color in biomedical research, as well as some criticism about their work. Although the first paragraph hinted at core elements of Burchard's work, it did so while perpetuating a damaging stereotype: the angry Latino. The article began:

> Esteban Burchard is pissed. He's angry that people of color are being sidelined by a medical revolution spawned by the Human Genome Project. He's angry that when non-white scientists apply for federal dollars to try and turn things around, they find the deck stacked against them. And he's angry that he, a middle-aged Latino man with a wrestler's build, can't walk his dogs without getting stopped by the police.[34]

Everyone should be angry about the injustices listed above. However, I was disappointed that Alhous—perhaps inadvertently—had leaned into stereotypes about Latinos. As we produced Burchard's trio of videos, we wanted to ensure that the story we told countered those stereotypes. In "Who's included?" Burchard provides several examples of how the sidelining of groups like women and African Americans has created health disparities and discusses how his lab is addressing them. He is passionate but never angry.

As a Latina, I am particularly aware of the "double bind" for minoritized women in science who experience simultaneous discriminations for intersecting identities.[35] Therefore, in "Charting an Original Path," Calisi Rodríguez's work as a researcher received as much attention as her advocacy for parents in science. To counter pervasive stereotypes of women researchers, such as their so-called lack of suitability for field research, we filmed her explaining her research in her lab and in aviaries where she keeps pigeons; Calisi Rodríguez's field research studying spiny lizards in Mexico as a master's student also received attention.[36]

We filmed the interviews using a documentary style, with all the scientists looking at the camera (interspersed with an occasional side shot), encouraging viewers to feel they are being directly engaged by the scientists. Poignant moments materialized through this approach, such as when Burchard unfolds a blanket-sized package insert for the asthma drug Serevent, adding:

> I wanna tell you why you don't read it. And it's because it's this huge road map of small text. But if you look here, it says that if you're African American and take this medication, you have a sevenfold-increased risk of dying.[37]

He then asks: "If you were a parent, would you let your child use this?"[38] By doing so, he appeals to viewers' empathy and challenges them to reflect on the shortcomings of a racially discriminatory healthcare system. Similarly, as Calisi Rodríguez recalls feeling isolated in graduate school because her identities were not welcomed in academia, viewers witness her struggling to find the right words to fully describe her experience. "So, the fact that— the fact that I came into a program with a background in art, as somebody who was very expressive and someone who is Mexican-American, I felt very alone. I felt very . . . I missed my community and that made it really hard and after my first year, I didn't think I would stay," she says.[39] These two moments from González Burchard and Calisi Rodríguez's videos allow viewers to experience a range of emotions—from shock to empathy—and connect with the protagonists.

From the start, we anticipated experimenting with different types of sounds and audiovisuals—from pictures to animations— to complement the interviews in B2B. We saw each film as an opportunity to try different styles that would help the stories shine. We obtained images from Burchard's childhood, his days as a wrestler, a medical student, and a physician resident, animating them as if they were being shown through an old projector (with corresponding sound effects) so that viewers felt they were journeying through his memories. For Calisi Rodríguez's film we relied more heavily on b-roll, following her at the 2018 Sacramento March for Science,[40] into her lab and aviaries, and at the farmer's market with her family. With Prakash we used secondary footage, too, opting to edit it in black and white; we interspersed color and black-and-white footage, an editorial choice made to visually contrast facets of his personality, like when he was enjoying assembling Foldscope or when he was sharing deep thoughts during the interview.

Animations offered a fun (and significant) creative challenge to me as a first-time short film producer. We incorporated them to explain complex or critical concepts, underscore pivotal narrative moments, and compensate for missing photographs or footage. In Burchard's "Which Box Do I Check?" he describes race as a "shopping cart that contains lots of information that is relevant for clinical and biomedical research." The accompanying animation depicts a shopping cart rolling through aisles of social determinants of health (e.g., poverty, discrimination), visually elaborating on his analogy. Animations are prominent in Prakash's video, helping to foreground the whimsy and wonder with which he approaches his research. Inspired by vibrant watercolors in different shades of red, yellow, blue, and green, the abstract animations feature different shapes and textures (dots, swirls, and waves) to add dimension.

Our videos made use of original music created by composers and material from music libraries as well as sound effects; the aural accompaniments helped to set the films' tones, amplify emotions, and signal important moments or pace changes. The ultimate goal of our filmmaking choices was to create and sustain multidimensional portrayals of scientists of color and marginalized backgrounds, thereby countering the limiting narratives typically constructed by media.

Encouraging Critical Conversations about Diversity, Equity, and Inclusion in Science

Given that one of the main goals of B2B is to challenge deficit narratives about minoritized scientists, we knew the films offered a great opportunity to encourage critical discussions about DEI in science. To this end, we have hosted screenings at academic institutions and public events. Generally, these consisted of showing some or all videos followed by a question-and-answer (Q&A) session with me and sometimes a production team member or one of the series' protagonists. The Q&As, guided by the interests of audiences, addressed ideas of racism in science, medicine, and academia; the scientists' research; DEI in biomedical research; the production process; next steps for the series; and suggestions for future films.

Our first screening happened at UCSF, soon after we released Burchard's videos in August 2017. The audience at this event was comprised mainly of faculty, staff, doctoral students, and postdoctoral scholars who worked there or at nearby institutions like Stanford or UC Berkeley. We showed all three videos, followed by a Q&A with Burchard, Griffin, and me. A simple survey captured the attendees' thoughts on the films. Respondents reported learning something new from the videos (forty-two out of fifty) and named "An Inclusive Future" their favorite video, although several said it was hard to choose just one.

Forty-four respondents said that they'd be interested in watching future videos in the series; many said they'd be willing to talk to others about the videos (thirty-three); look for more information on the topics discussed (twenty-one); share the videos on social media (nineteen); and watch the videos again (sixteen). Based on the comments, respondents enjoyed the animations and felt that all three of Burchard's videos helped elaborate on the complexities of racism in medical science and the importance of inclusion. His personal stories also resonated with attendees; one respondent, for example, recalled having professors who deemed applicants "not culturally ready" for particular postdoctoral positions. Attendees also appreciated the opportunity to interact with Burchard, the editor, and me as part of the screening.

After the positive feedback we received at UCSF, we wanted to assess if B2B could change perceptions about belonging in science as well as how someone's cultural and ethnic background contributes to success in science. To answer these questions, in collaboration with colleagues from the National Research Mentoring Network (NRMN) at the University of Wisconsin-Madison, we developed and distributed a different survey[41] after screenings of Burchard's videos at SFSU and Guttman Community College in New York City.[42] Of the forty-five people who responded to this survey, twenty-nine were undergraduate students, fifteen were master's students, and one was a professor. Most of them identified as non-white (thirty out of forty responses), including Hispanic/Latino, Black/African American, Asian, Native Hawaiian or Pacific Islander, American Indian or Alaska Native, and Other. Most people (twenty-nine out of thirty-nine responses) said that B2B videos were extremely or moderately valuable in helping them understand how one's ethnic and cultural background can contribute to a successful career in science. On a scale of 1 to 5 (none to a lot), respondents reported the videos improved their understanding of the value of diversity in scientific research (mean 4.05 out of 5), and twenty-seven out of thirty-nine said they were very likely or likely to recommend the films to someone else.[43]

The comments of SFSU and Guttman survey respondents indicated that Burchard's anecdotes and experiences resonated with them. Of the video "Which Box Do I Check?" one commenter noted, "I like this video because it exposes the complexity surrounding race, ethnicity, and ancestry. I never know which box to check, and most of the time they don't have any of my boxes." Some reported learning about the importance of perseverance and the impact biases can have in people's health from watching Burchard's series. "Changing the story to showcase how our diverse experiences can be used to our benefit in our own research," wrote a respondent. "I had heard of this idea, but hadn't been given a strong example until watching this video."[44] Importantly, the videos raised concerns about how studying genetic differences between racial and ethnic groups can be used to discriminate against, not help, marginalized populations, a common criticism Burchard receives from other scholars. As one respondent asserted: "My only concern is, a conservative racists [sic] could interpret this as 'yes there are genetic differences between races and sex's [sic]' in a negative sense and use it to preach their mean ideas."[45]

In addition to the abovementioned events, we have hosted screenings of all B2B films at the Bay Area Science Festival and the Society for Advancement of Chicanos, Hispanics, and Native Americans in Science (SACNAS) Annual Conference, among others. I have also presented B2B videos as case studies for how films can help change science stereotypes and advance science communication at institutions and conferences across the United States.[46] These presentations offered opportunities to articulate and present the conceptualizing, creating, and releasing of B2B as Practice as Research. For example, as part of a talk at the 2019 National Academy of Sciences Colloquium held in Irvine, California,[47] I recalled how during editing, executive producer Kirschner said to me and the editor: "Let's take a step back and think about how he is explaining his research . . . where is he lighting up?" He challenged us to go back to the editing board and create a final version where viewers could connect with Prakash's emotions, his excitement, and sense of wonder and curiosity, which the earlier version of the video failed to do. During my talk, I showed different versions of Prakash's video, one in which he seemed subdued and used quite a bit of jargon to explain his research and a second in which he was more animated and explained his work more accessibly. By contrasting these versions, I showed how we used different takes, filmmaking techniques (e.g., animations), and editorial choices (e.g., making b-roll black and white, detailed earlier) to challenge old ideas about scientists while innovating in the communication of science through film.

Digital multimedia—and more broadly, communication—are vital tools for fostering critical discussions about representation, diversity, equity, and inclusion in science and beyond. We at the Science Communication Lab created "Background to Breakthrough" to challenge deficit narratives about scientists of color and from marginalized backgrounds. We saw the films as an opportunity to innovate in filmmaking, using different techniques to reframe and counter one-dimensional portrayals of minoritized scientists. Responses to the videos online, at events, and through surveys suggest that we have been successful.

Notes

1 Cary Funk, Jeffrey Gottfried, and Amy Mitchell, "Science News and Information Today," *Pew Research Center's Journalism Project*, 20 September 2017, https://www.pewresearch.org/journalism/2017/09/20/science-news-and-information-today/.

2 Ibid.

3 Naomi Forman-Katz and Katerina Eva Matsa, "News Platform Fact Sheet," *Pew Research Center*, 20 September 2022, https://www.pewresearch.org/journalism/fact-sheet/news-platform-fact-sheet/.

4 Jeffrey N. Schinske, Heather Perkins, Amanda Snyder, and Mary Wyer, "Scientist Spotlight Homework Assignments Shift Students' Stereotypes of Scientists and Enhance Science Identity in a Diverse Introductory Science Class," *CBE—Life Sciences Education* 15, no. 3 (2016): 47, https://doi.org/10.1187/cbe.16-01-0002.

5 "Portray Her: Representations of Women STEM Characters in Media," *Geena Davis Institute*, 9–10, https://seejane.org/research-informs-empowers/portray-her/.

6 Christopher Lloyd, *Back to the Future* (1985, Universal Pictures); Paul Zaloom, *Beakman's World* (1992–7, Columbia Pictures Television).

7 Johnny Galecki, *The Big Bang Theory* (2006–19, Warner Bros. Television).

8 "Portray Her," 4.

9 Cary Funk, "Black Americans' Views of and Engagement with Science," *Pew Research Center Science & Society*, 7 April 2022, https://www.pewresearch.org/science/2022/04/07/black-americans-views-of-and-engagement-with-science/; Cary Funk and Mark Hugo Lopez, "Hispanic Americans' Trust in and Engagement with Science," *Pew Research Center Science & Society*, 14 June 2022, https://www.pewresearch.org/science/2022/06/14/hispanic-americans-trust-in-and-engagement-with-science/.

10 Monica L. Ridgeway, "Deprogramming Deficit: A Narrative of a Developing Black Critical STEM Education Researcher," *Communications on Stochastic Analysis* 18, no. 1 (2019), https://doi.org/10.31390/taboo.18.1.11.

11 Ursula Nguyen and Catherine Riegle-Crumb, "Who Is a Scientist? The Relationship between Counter-Stereotypical Beliefs about Scientists and the STEM Major Intentions of Black and Latinx Male and Female Students," *International Journal of STEM Education* 8, no. 1 (2021): 28, https://doi.org/10.1186/s40594-021-00288-x. See also, Schinske et al.

12 "The Wonder Collaborative: Reinventing Science Filmmaking," WonderCollaborative, wondercollaborative.org.

13 SCL is a nonprofit organization dedicated to using multimedia storytelling to engage the public, including educational and scientific communities, in the journey and wonder of science. It can be found at https://www.sciencecommunicationlab.org/.

14 "Background to Breakthrough," *iBiology*, https://www.ibiology.org/stories/background-to-breakthrough/. B2B is based upon work supported by the National Institute of General Medical Sciences (NIGMS). The videos were produced in collaboration with the National Research Mentoring Network (NRMN), which is also funded by the National Institute of Health (NIH).

15 Robin Nelson, *Practice as Research in the Arts and Beyond* (London: Palgrave Macmillan, 2022), 40.

16 Wonder Collaborative, "Background to Breakthrough Playlist," *YouTube*, https://youtube.com/playlist?list=PLENwi7txclRBmjSHRR28yJbO2mXWKKT2S.

17 *Part 1: Racial Bias in Science and Medicine: Who's Included?*, Esteban Burchard, iBiology, 2017, documentary film, 4:22, https://www.ibiology/org.background-to-breakthrough/improving-diversity-inclusion-science/#part-1.

18 Arjun K. Manrai et al., "Genetic Misdiagnoses and the Potential for Health Disparities," *New England Journal of Medicine* 375, no. 7 (2016): 655–65, https://doi.org/10.1056/NEJMsa1507092.

19 *Part 2: The Impact of Race and Genetic Ancestry on Medicine: Which Box Do I Check?*, Esteban Burchard, iBiology, 2017, documentary film, 4:13, https://www.ibiology.org/background-to-breakthrough/improving-diversity-inclusion-science/#part-2.

20 *Part 3: Inclusion of Minorities in Science and Medicine: An Inclusive Future*, Esteban Burchard, iBiology, 2017, documentary film, 6:49, https://www.ibiology.org/background-to-breakthrough/improving-diversity-inclusion-science/#part-3.

21 *Charting an Original Path*, Rebecca Calisi Rodríguez, iBiology, 2018, documentary film, 10:15, ibiology.org/background-to-breakthrough/charting-an-original-path/.

22 Sarah Colwell, "Fusion of Science and Parenting is Focus of Professor's Research, Advocacy and Art," *University of California Davis, College of Biological Science*, 26 April 2022, https://biology.ucdavis.edu/news/authenticity-and-public-scholarship.

23 "Pigeon Milk Is a Nutritious Treat for Chicks," *Audubon*, 11 March 2019, https://www.audubon.org/news/pigeon-milk-nutritious-treat-chicks.

24 Rebecca Calisi Rodríguez (@BeccaCalisi), *Hold up #SFN17. I'm not onsite yet, but I was just informed THESE are your so called lactation/childcare rooms?* Twitter, 10 November 2017, https://twitter.com/BeccaCalisi/status/929009552028180480?s=20&t=sAuaRAxNmllojNuypPwWtQ.

25 Rebecca Calisi Rodríguez (@BeccaCalisi), *I've come to find out #SfN17 actually DOES NOT provide adequate resources for breastfeeding/pumping mothers.* Twitter, 10 November 2017, https://twitter.com/BeccaCalisi/status/929051203782066176?s=20&t=4tenWMmJi4y1btQmTES_Wg.

26 Rebecca Calisi Rodríguez (@BeccaCalisi), *Looks like #SFN17 has stepped it up! Childcare/lactation room in 204A now has more comfy chairs available.* Twitter, 13 November 2017, https://twitter.com/BeccaCalisi/status/930131825980305410.

27 Rebecca M. Calisi, Working Group of Mothers in Science, "How to Tackle the Childcare-Conference Conundrum," *Proceedings of the National Academy of Sciences* 115, no. 12 (2018): 2845–9, https://www.pnas.org/doi/abs/10.1073/pnas.1803153115.

28 *Scientific Curiosity: Finding Sublime in the Mundane*, Manu Prakash, iBiology, 2018, documentary film, 8:47, http://www.ibiology.org/background-to-breakthrough/scientific-curiosity/.

29 *Prakash Lab: Curiosity-Driven Science*, "Frugal Science and Global Health, https://web.stanford.edu/group/prakash-lab/cgi-bin/labsite/research/frugal-science-and-global-health/.

30 *Scientific Curiosity*.

31 Prakash Lab, Abuzz: citizen-based mosquito monitoring system, https://web.stanford.edu/group/prakash-lab/cgi-bin/mosquitofreq/the-science/.

32 Foldschope Instruments, Inc., "The Paper Microscope," *Foldscope Instruments*, foldscope.com.

33 Manu Prakash, *A 50-cent microscope that folds like origami*, TED: Ideas worth spreading, 2012, TEDtalk, 8:23, https://ted.

com/talks/manu_prakash_a_50_cent_microscope_that_folds_like_origami?language+en.

34 Peter Alhous, "Meet The Scientists Fighting for More Studies on Genes and Racial Differences in Health," *Buzzfeed*, 11 May 2016, https://www.buzzfeednews.com/article/peteraldhous/science-so-white.

35 Shirley Malcom, Paula Quick Hall, and Janet Welsh Brown, "The Double Bind: The Price of Being a Minority Woman in Science," *American Association for the Advancement of Science* (1976): 2.

36 Rebecca M. Calisi and Diana K. Hews, "Steroid Correlates of Multiple Color Traits in the Spiny Lizard, *Sceloporus Pyrocephalus*," *Journal of Comparative Physiology* 177 (2007): 641–54.

37 *Racial Bias in Science and Medicine*.

38 Ibid.

39 *Charting an Original Path*.

40 March for Science is a community of science advocates that began as an international series of rallies and marches held on Earth Day, 22 April 2017, in Washington, DC, and in more than 600 other cities across the world. Since then, yearly marches have been held.

41 Kim Spencer, Christine Pfund, and Mónica I. Feliú-Mójer. Online survey. September–November, 2018.

42 We only surveyed viewers about Burchard's videos because we had not yet finished production of Calisi Rodríguez and Prakash.

43 Ibid.

44 Ibid.

45 Ibid.

46 A partial list of these includes the University of North Carolina, University of Colorado, University of California, Davis, Max Planck Institute for Biophysical Chemistry, Society for Neuroscience Conference, the Science Communication Colloquia of the National Academies of Science, Santa Rosa Junior College, Science Talk Conference, Conference, Jackson Wild Film Festival, University of Oregon, and Cultivating Ensembles. You can read more about Cultivating Ensembles in Raquell Holmes's essay in Volume 1 of this collection: Raquell Holmes, "Cultivating Ensembles: A Relational Reflection on Creating Cultural Transformation with New Performances of Science," in *Identity, Culture, and the Science Performance, Volume 1: From the Lab to the Streets*, ed. Vivian Appler and Meredith Conti (London: Methuen Drama, 2022), 54–71.

47 Mónica Feliú-Mójer, *Monica Feliu Mojer - Telling Stories to Change Misconceptions About Scientists*, NAS Colloquia, 2019, lecture, 8:09, https://youtube.com/watch?v=m0p4SrReq_Q.

6

Celestial Politics

Performance and the Cosmic Underclass

Felipe Cervera

In the introduction to the first volume of this collection, Vivian Appler and Meredith Conti highlight that "while performance has itself sustained racially, culturally, and financially exclusive systems and practices, it is also an instrument for intervention and revolution [. . .] at the intersections of science and performance."[1] In this chapter, I make the case that the definition of the science performance and its politics can also be extended to instances where the science is hidden from the critical view, but perhaps more importantly, hidden in the folds of what makes a science performance scientific or political in the first place. To advance this point, I propose the notions of the cosmic underclass and the celestial politics of performance. I argue that cultural critique must

---

This chapter evolves from a keynote titled "Tropical Sovereignty: Performances of Resistance and Alliances" that was presented during the conference Tropical Performances: The Second Performance Studies Philippines Conference, hosted by De La Salle University, in Manila, the Philippines, on August 5–6, 2019.

pay attention to these as most individuals may be so busy trying to ascertain their right to exist on this planet's soil that their political and artistic expressions may only *imply* their concurrent claim to live under a sky of their own. If, as performance scholar Lynette Hunter argues, the goal of any political performance—understood broadly from public political demonstrations to private dance or theatre pieces—is to unsettle the certainties of governance at any given time,[2] then the certainties to be challenged should include those we hold about the sky. Cultural astronomer Giulio Magli argues that "[i]ndividuating the Cosmos[...]and opening up communication between cosmic levels are[...]fundamental operations for human societies. The performance of such operations, consequently, yields power."[3] If this is the case, performance studies must develop methodological and critical tools to unearth the sky from the epistemological erasure that impedes some peoples on this planet from enacting their celestial politics. A productive way to open this conversation is to look at examples of performances that contest and unsettle the stability of identities as they are shaped and enacted by and through our relation to land to establish then the continuity among mythical, symbolic, and metaphorical uses of the sky; its consequent semantics in performance; and the material occupation of suborbital space and beyond. The material relationship of life with the planet is such that the disassociation of land, water, and sky when we think of political agency is limiting and inaccurate. Our analysis must extend to how *the land is used to access the sky and define its politics.*

A Point on Methodology: The Cosmic Underclass

Sociologists Peter Dickens and James Ormrod propose the notion of a "cosmic elite" and suggest that this is a constant across different societies.[4] In their study, Dickens and Ormrod trace a history of the cosmic elite across different periods of European and Eurocentric histories. Along the way, they emphasize how colonial transoceanic travel after the medieval European period is also marked by an epistemic change in astronomy and, consequently, in the ruling cosmic elite. Cosmic elites ceased focusing on validating religious and absolute power on earth through celestial geometries to become instead invested in developing astronomy toward the goal of colonizing the planet.[5] Key to understanding what this transition means is to see that the cosmic elite fundamentally implies a division of labor that is "orchestrated" and is a "basis for class and other social divisions" closely aligned with colonial strata.[6]

In their proposal of using "the cosmic elite" as a methodological focus to trace shifts in astronomical (and later astronautical) knowledge and power, Dickens and Ormrod also emphasize that "the nature of the cosmic elite may change over time, but its power does not necessary recede."[7] This is important because while their power and agency may remain constant, their practices (or performances, as Magli calls them) to hold this power may change. The origins of cosmological empiricism may be traced to proto-scientific practices in societies like the Greek. Later, its preeminence as the ruling episteme becomes apparent with the rise of modern astronomy in the science performance of Copernicus, Kepler, Galileo, and their successors before becoming hegemonic in the twentieth century with the rise of astronautical practices.[8] Today, the cosmic elite "are no longer just priests, shamans, and astrologers (though, along with imams and the astrological elite of the world religions, they continue to play a role), but astronomers, astrophysicists, aerospace engineers, astronauts, and key figures in the civilian, military, and corporate space programs."[9] Evidently, the argument here is not that these groups of people are necessarily aspiring to some malicious control of the sky, but rather that their science performance determines much of how the rest of the people on this planet defines their relationship with it.

The work by Dickens and Ormrod is pivotal to elaborating a more thorough critique of scientific power (and indeed science performance) in the twenty-first century. However, their analysis still needs to be expanded to understand the epistemological and political implications of the evolution of cosmic elites outside their immediate field of action and influence. And while they do suggest that an obvious consequence of the cosmic elite is generating what I am calling here a cosmic underclass, they do not go on to define what this might be or whether it is valuable to study it beyond astronomical contexts. I see in this a methodological shortcoming, and to remedy it I propose to define the cosmic underclass and identify it beyond astronomical and astronautical practices. This group includes those who cannot see the sky above them as directly impacting their political circumstance on land, those for whom the politics of their sky remain separate from the politics of the ground underneath their feet. Today's cosmic underclass is most of us for whom our access to the sky is entirely mediated by layers of complicated networks of political and scientific actors who determine how and why the sky matters to us.

I propose exploring the dynamic relationship between these two poles of spectrum is necessary because thinking *only* about the agency cosmic elites might have on the land they use to access the sky leaves us in the middle of an epistemic bind. This is, while we can map how the sky is being produced as a social, cultural, and political atmosphere for and by them, we might not be able to detect how the *cosmic underclass* is being affected by and posing challenges to the celestial politics enacted by the elite insofar as astronomical knowledge and power are exclusive to those who have and perform access to the sky. The methodological challenge in elucidating the cosmic underclass is, therefore, that their agency and expression might not be considered as initially relevant to discussions about celestial politics, as they may remain invisible from any analysis that pays attention to the power dynamics between two or more cosmic elites and their science, ritual, or artistic performances. The underclass's actions might not be perceptible enough to redistribute the sensible, as philosopher Jacques Rancière might have it.[10] Or, following postcolonial thinker Gayatri Spivak's critical vocabularies, they might be erased from scientific historiography due to their subalternity.[11] In other words, in the attempt to identify and render justice to the cosmic underclass, there might be the reflex of finding an example that quite literally takes place in the sky or that in some way engages directly and perceptibly with the access to or representation of the sky. Identifying the cosmic underclass is different from the work of identifying previous iterations of cosmic elites who remain perceptible as such (such as Indigenous elders who perform ritual access to the sky) even after they have lost the monopoly of the sky due to settler colonialism or technoscience.

An excellent example of a conflict between cosmic elites is the case of the protests in Hawai'i against the construction of a Thirty-Meter-Wide Telescope (TMT).[12] The TMT is meant to be built on the peak of Mauna Kea as part of the more extensive system of observatories already in place there and will be managed by a consortium of universities in the United States, Canada, China, India, and Japan. It is designed as a giant mirror interacting with hypersensitive photo and radio sensors to access the deep universe at a resolution twelve times higher than those produced by the Hubble telescope and at least four times sharper than the ones made by the new James Webb Telescope, both of which are placed outside of Earth's atmosphere. The rationale for wanting to build the telescope there is that the height of Mauna Kea and the clarity of the night sky offer the ideal location for this kind of instrument. However, Hawai'ian activists have been protesting the telescope since it was first proposed in 2015, not because of the telescope itself but because Mauna Kea is a sacred place in their ritual geography. Each of these elites is defending a distinct cosmic agency and subjectivity, and neither claims to nullify the other, but both are contesting the land on which to access the sky. This is a fascinating debate. But simply focusing on the competition that cosmic elites, past and present, may have on the land on which they need to stand to perform celestial access is only half the work. We still need to account for how the agency of these various overlapping cosmic elites shapes the subjectification of the cosmic underclass. This stratification of power reflects the interplays between the politics of the sky and the politics of the land and determines the few who can see and be seen by the sky.[13]

To further my case, I rely on work carried out by feminist, postcolonial, and decolonial theoreticians who have argued for dismantling structures of power and knowledge that erase epistemologies by stratifying knowledge across vertical axes. In specific, I rely on Sandra Harding's work who, on her way of articulating feminist science and technology studies "from below," argues that to disassemble problematic binaries such as First/Third World, West/Rest, European/non-European, industrialized/nonindustrialized, we must first signal how problematic these binaries are. Harding argues we should avoid "[m]oving away from that binary before fully appreciating how its persisting legacy falls more under the category of racist and imperial denial than of a progressive project contributing to the dismantling of that legacy."[14] The occupation of land by global powers to access the sky is a colonial legacy that requires careful attention in the project of negotiating and rendering justice to different epistemological traditions, given that one of its most evident consequences is the ontological stratifications that articulate white and androcentric technoepistemologies as being above everyone else's knowledge and experience of the planet, quite literally. I thus echo Harding in suggesting we must find a way to enable discussions "from below" about how land-based cultural practices may unsettle the monopolization of the skies currently underway.

I propose that articulating performance-based methodologies to unearth the agency of the cosmic underclass follows the spirit of understanding, first, how the binary elite/underclass is enacted and perpetuated to then challenge that epistemological bias through the inclusion of and attention to practices that previously may not have been deemed relevant to the debate. Here, I am building on work already done by performance studies scholars on what we call critical outer space studies or, perhaps more simply, the anthropology of outer space.[15] I build on my own work in "Astroaesthetics: Performance and the Rise of Interplanetary Culture,"[16] where I alerted that the performativity of astronautical practice has, since the 1950s, advanced an androcentric colonialist narrative onto the solar system and highlighted the possibilities found on the work done by sound artists who, in using radio astronomical data as media, offer alternative planetary epistemologies. I also build on the work done by Appler in her "Titan's 'Goodbye Kiss': Legacy Rockets and the Conquest of Space," where she draws from the work of Diana Taylor on the notion of a conquest scenario to chiefly identify that "the extraterritorial nomenclature system, as performance in the *Gazetteer [of Planetary Nomenclature]* can be traced to the cosmographical processes deployed during the naming of the Americas."[17] Lastly, the argument here has resonances with the work done by Scott Magelssen in *Performing Flight: From Barnstormers to Space Tourism*,[18] where he highlights the white male supremacist lineage of figures like the astronaut and the space tourist/entrepreneur.

A central focus of this chapter is therefore to broaden the critical entry points that performance studies can offer to the interdisciplinary study of outer space. Indeed, as a field on the periphery of astronomy and astronautics, performance studies' contributions to the critical study of outer space have mostly been centered on the analysis and conceptualization of performances that unsettle the politics of hegemonic cosmic elites. However, precisely because much more of the work that performance studies does is focused on the politics of performance *on land*, we are in a prime position to open a conversation that avoids leaving the debate about the sky to the very exclusive group of artists and scholars who directly interact with and have access to astronomical and astronautical institutions and data. To achieve this, I argue we can explore how performances may express and articulate the celestial politics of performances that are not evidently about the sky.

Celestial Politics in the Work of Lukas Avendaño

Lukas Avendaño is a Zapotec *muxe* performance artist and anthropologist. Zapotecs are one of the largest Indigenous nations in Mexico, with a current population of close to one million, mainly located in the state of Oaxaca, in the southeast of Mexico. They recognize three genders: male, female, and *muxe*. However, it would be erroneous to assume that these three genders exist in the Zapotec context as stable gender categories and identities. The Zapotec language does not distinguish between male and female, and the process of gender identification is much more performative than in a context where the gender categories are firmly defined by language and its socialized embodiment. This gender fluidity predates the arrival of Spanish colonialism in Mexico. It has since gone through obvious complications during the colonial and postcolonial periods, especially after conservative religious values were enforced as repressive mechanisms against Indigenous nations' identities and social norms.[19]

Muxes' complicated history is the starting point from which Avendaño troubles the stability of the Mexican identity and the genders it recognizes.[20] Avendaño's work illuminates the invisible colonial legacies of gender codes among Indigenous communities in Mexico and the systematic oppression and structural violence exercised onto him by the postcolonial and modern Mexican state.[21] To do this, Avendaño activates his *muxe* identity and Indigenous gender politics to interrogate how he becomes visible or invisible in the Mexican context. Avendaño fleshes out the embodied traces of precolonial, colonial, and postcolonial identities in what we can perceive as part of the Indigenous critique of Mexican modernity, its syncretism, and the implications its laws and norms have for Indigenous communities.

A key work in Avendaño's *oeuvre* is *Requiem para un Alcaraván* (Requiem for a Stone-curlew).[22] The work is a series of scenes in which Avendaño enacts female rites of passage typical of the Zapotec communities in the Tehuantepec Isthmus. Avendaño performs as a bride, as the organizer of the feast for the local patron saint, as a healer, and as the host of a funeral before concluding by performing, alone, the traditional dance of the curlew, which is a pre-Hispanic love dance in which the dancers mimic the mating of these birds. Antonio Prieto is a leading scholar of Mexican performance studies and longtime student of Avendaño's *oeuvre*. He argues that *Requiem* is a performative intervention in the imaginary of the *tehuana* (women of Tehuantepec). Prieto alerts how *Requiem* makes a poignant commentary on how Indigenous women in general and the *tehuana* in specific are abused as central iconographies in the postcolonial narrative of modern Mexico. The piece messes with national discursivity that frames the Zapotec woman as a central reference to the ancestral heritage that modern Mexico is proud to uphold by inserting Avendaño's *muxeidad* in its performance. In doing so, Prieto argues, Avendaño challenges the objectification that the Indigenous gendered body is subject to as a strategy for national branding even when Indigenous women remain so largely disempowered. This tendency has marked Mexican discursivity since the Revolution of 1910, when, as Zuzana M. Pick identifies, the *tehuana* became a "gendered icon of authenticity in the practices of *mexicaneidad* (Mexicanity)."[23] Indeed, Frida Kahlo used the *tehuana* as a staple image in her work, making it a visual reference *de rigour* of Mexican modernity and, more recently, the icon of global Mexican femininity. Prieto thus observes that Avendaño's *Requiem* can be taken as a performative intervention into the discursivity of Indigenous gender in Mexico.

The second layer of *Requiem* is about land and the right to be on it. The scenes in the piece are also part of the repertoire of dances/performances in the Guelaguetza Festival, one of the Indigenous festivities in Mexico par excellence. It takes place during the final week of July of every year and has its own venue—the *Guelaguetzodromo*. It is a significant touristic attraction in Oaxaca as well as a major income producer for the tourism industry at a national level. The Guelaguetza is heavily codified in terms of gender, distribution of labor, and cultural identity of the different ethnic groups that gather there. The festival holds a sociopolitical meaning within the alleged ancestral norms that governed and continue to govern the cohabitation of the other nations and Indigenous groups that share territories in Oaxaca and nearby. The social gesture performed during the event is the gifting of produce grown by every region to maintain their bonds. It is a performance of generosity and, above anything else, of political alliance with and through the land. Yet, *muxes* have been deterred from participating because outside of Juchitan, their identity means trouble. As Avendaño often says, "one is *muxe* in the Isthmus, but outside it, and to the eyes of the other, I am a *puto* [faggot]/homosexual or whatever other similar denomination."[24]

Requiem can be interpreted as an expression of Avendaño's frustration with not being able to perform these rites, if not during a performance situation or inside his community, precisely because of the impossibility attached to their gendered indigeneity. The piece is thus a performance of political contestation and resistance that finds much of its power in interrogating the relationship between land and body, thus making the ghost of sovereignty that lingers on the heads of the Indigenous communities in Mexico visible. Indeed, as the 2021 report from the National Committee for Human Rights (Comisión Nacional para Los Derechos Humanos) highlights, despite several modifications to the country's Constitution and legal framework that recognize Mexico as a plurinational state, Mexican Indigenous nations remain in a situation of structural discrimination and therefore are unprotected and disavowed as agents of their own political will, destiny, representation, and right to live and be on their ancestral land according to their self-determination.[25]

Avendaño's treatment of gender and indigeneity establishes the groundwork for a nuanced discussion of celestial politics. The above is the now-standard analysis that several scholars, including Prieto, have done about their work.[26] So far, no attention has been given to Avendaño's implicit and explicit references to the sky and celestial politics. But the sky is very much underneath Avendaño's feet as he performs *Requiem*, and limiting the analysis to the claims the performance makes to the land without even paying attention to the relationships it implies to the sky would overlook the importance of addressing the possibility of celestial politics in the work.

A closer look at the sky's symbolic and representational presence in Avendaño's piece will help us trace the extent to which the sky remains so extremely colonized that it is taken for granted. Excavating precolonial identities reveals the connections that they might imply between land, body, and sky. "Zapotec" is not the name by which Zapotecs call themselves but is what the Nahuas used to call them. "Zapotec" is derived from the Nahuatl *tzapotecah*, which means "the people of the land of sapote" and *sapote* is the fruit they paid as taxes to the Nahuas.[27] The name remained once Spain colonized Mexico-Tenochtitlan because the names Nahuas used to call the rest of the nations under their power remained operational as the Spanish colonial state was erected. *Muxe*, too, is an imposition, as it is a word that could derive from the Spanish *mujer* (woman).[28] The people from the place of *sapote*; the not-not-women from the site of *sapote*.

The name that Zapotecs give to themselves, however, reveals a nuance that may not seem very important in an analysis of Avendaño's work at first sight but is nevertheless vital toward the thesis of the cosmic underclass. The actual name of the Zapotecs is *Ben 'Zaa* or *Binnizá*, which means "the people from the clouds" in their language, which is called *Diidx zah*.[29] There is a substantial epistemological and historiographic difference between being known and written as the people named after a tax paid to an imperial state on the one hand and being known under the self-determining name that connects one's collective subject with the sky and its cultural histories on the other. Cognitive philosopher Shawn Gallagher reminds us that naming is an institutional practice that creates invisible yet powerful legacies.[30] The politics in the genealogy of naming the Zapotecs as such are crucial to understanding the system of forced trade that underlined the social construction of indigeneity in precolonial and early colonial Mexico and the consequences that this had on the severance of cultural relationships between land, identity, and the celestial realm. Indeed, the politics of the sky start with the power to name one's own relationship to it.

Following that thread, the presence of celestial power itself appears more clearly in *Requiem*. My scope here encompasses the symbolic presence of the sky in the figurative sense, especially regarding religion. Avendaño's piece is dotted with references to Catholic rites and prayers, all of which were instituted in Mexico, and throughout much of the Americas, via theatrical representations and as a cornerstone of a colonial project that aimed not only to take the land and submit the people, but also to occupy its sky (and underworld) with Catholic iconography and the power it brings. The first Catholic evangelical play to ever be staged in Mexico (sometime between 1531 and 1548) was titled *The Final Judgement*. It was as powerful an instrument of celestial colonization as it was in enforcing gender binaries and repression against Indigenous women. The play's author is unknown. However, it is usually attributed to Andrés de Olmos, a Franciscan Friar who was part of the first evangelical missions in Mesoamerica. The play was staged in Nahuatl, and its story represents the final judgment of Lucia, an Indigenous woman who has sex with multiple partners. Lucia is directly punished by Jesus Christ, who descends from heaven for the occasion.[31] Elsewhere, I have referred to the play and its representation to explain the eschatological politics of theatre, claiming that the characterizations and representations of the end of times, its agents, the afterlife, and the powers that control them may be weaponized on the grounds of creating a sense of apocalyptic agency.[32] This apocalyptic agency is, at least in Mexico and Tehuantepec, very closely connected with the symbolic occupation of the sky as the place where Christian redemption or punishment may be found. Indeed, when Avendaño ends their *Requiem* with a series of supplications that include asking the "father celestial" to forgive all their sins, they not only actualize the cultural syncretism that shaped Mexican modernity in general and religion in specific, but they also *imply* the impossibility of the *Ben 'Zai* and *muxes* to imagine a sky of their own—or, at least, one in which they have material and symbolic agency—a sky that gives them the power to be whom they want to be on their land.

Conclusion

My goal in this chapter has been to make a methodological argument: celestial politics may already be present in any form of political performance that challenges the right to occupy the land, regardless of whether it explicitly references the sky. The extent of a decolonial project cannot be limited to the successful return of occupied land, as Eve Tuck and K. Wayne Yang have argued,[33] but it should also include the successful reopening of the skies. In Avendaño's work, the various ways the sky is embedded into and expressed as performance are essential. They imply a sky hidden from the view of those who may not readily see themselves as agents of celestial power and dynamics—that is, those who belong to the cosmic underclass. Is *Requiem* about gender in space exploration? No. Is *Requiem* about the cultural conflict between pre-Hispanic and colonial ontologies of the sky and the universe? No. But does *Requiem* imply these points? Yes. If the analysis of the land-oriented, or terrestrial, politics of Avendaño's work is separated from the celestial politics that inform and shape its context, we miss an excellent opportunity to enable a cultural critique that makes the implicit politics of the sky explicit. Echoing Harding, we would miss the chance to formulate a critique of the scientific construction of outer space *from below*: we would miss making productive connections between decolonial theory and performance practice with the anthropology of outer space and its critiques. Recalling and reframing the name of the people that Avendaño belongs to is not just another anthropological footnote that relates to some mythical folkloric past, but is also a critical reminder that cultural histories of the sky that predate, parallel, and follow the advent of space flight exist.

Notes

1 Vivian Appler and Meredith Conti, "Introduction: Taking It to the Streets: Performing Science in Public," in *Identity, Culture and the Science Performance Volume 1: From the Lab to the Streets*, eds. Vivian Appler and Meredith Conti (London: Bloomsbury, 2023), 5.

2 Lynette Hunter, *Politics of Practice: A Rhetoric of Performativity* (London: Palgrave Macmillan, 2019), 15.

3 Giulio Magli, *Archaeoastronomy: Introduction to the Science of Stars and Stones* (Cham: Springer, 2016), 80.

4 Peter Dickens and James S. Ormrod, *The Cosmic Society: Towards a Sociology of the Universe* (London: Routledge, 2007), 4.

5 Ibid., 32.

6 Ibid., 21.

7 Ibid., 18.

8 Ibid., 2.
9 Ibid., 4.
10 Jacques Rancière, *The Politics of Aesthetics: The Distribution of the Sensible* (New York: Continuum, 2004).
11 Gayatri Chakravorty Spivak, "Can the Subaltern Speak?" in *Marxism and the Interpretation of Culture*, eds. Cary Nelson and Lawrence Grossberg (Chicago: University of Illinois Press, 1988).
12 More on the protests at Mauna Kea can be found in Māhealana Ahia, "Mālama Mauna: An Ethics of Care Culture and Kuleana," *Biography* 43, no. 3 (2020): 607–612, doi:10.1353/bio.2020.0068. For more on the TMT, Mauna Kea Observatories, and Hawaiian astronomy from a performance studies perspective, see Vivian Appler and Kenya Gadsden, "The 'A' in STEAM: PAR as Fifth-Space for Research and Learning in the Arts and Sciences," *PARtake: The Journal of Performance as Research* 4, no. 1 (August 2021), doi: https://doi.org/10.33011/partake.v4i1.533.
13 For more on the protests surrounding the TMT, see Alexandra Witze, "Hawaiian Telescope Project Sparks Protests at Astronomy Meeting," *Nature* (2015), https://doi.org/10.1038/nature.2015.18125. For more on Hawaiian cosmology, see: Naleen Naupaka Andrade and Cathy Kaheau'ilani Bell, "The Hawaiians," in *People and Cultures of Hawai'i: The Evolution of Culture and Ethnicity*, eds. John F. McDermott and Naleen Naupaka Andrade (Honolulu: University of Hawai'i Press, 2011).
14 Sandra Harding, *Sciences from Below: Feminisms, Postcolonialities, and Modernities* (Durham: Duke University Press, 2008), 157.
15 There is a growing interest in the interdisciplinary and critical study of outer space. Key publications include *Securing Outer Space: International Relations Theory and the Politics of Outer Space*, eds. Natalie Bormann and Michael Sheehan (London: Routledge, 2009) and *The Palgrave Handbook of Society, Culture and Outer Space*, eds. Peter Dickens and James Ormrod (London: Palgrave, 2016).
16 Felipe Cervera, "Astroaesthetics: Performance and the Rise of Interplanetary Culture," *Theatre Research International* 41, no. 3 (2016): 258–75.
17 Vivian Appler, "Titan's 'Goodbye Kiss': Legacy Rockets and the Conquest of Space," *Global Performance Studies* 2, no. 2 (2019), doi: https://doi.org/10.33303/gpsv2n2a5.
18 Scott Magelssen, *Performing Flight: From the Barnstormers to Space Tourism* (Ann Arbor: University of Michigan Press, 2020).

19 For more on gender and identity norms in colonial Mexico, see Steve J. Stern, *The Secret Story of Gender: Women, Men & Power in Late Colonial Mexico* (Chapel Hill: The University of North Carolina Press, 1995).

20 Guillermina Bevacqua, "Devenir muxe: torsiones desobedientes de Lukas Avendaño en *Réquiem para un alcaraván* y en *Buscando a Bruno*," *L'Ordinaire des Amériques* 228, 11 March 2022.

21 Avendaño identifies as *muxe*, which, as described here, does not directly translate with Euro-colonial gender categories, including transgender. In line with most scholarship published by and about Avendaño, this essay refers to the artist using masculine gender pronouns.

22 To watch a full documentation of the performance visit: https://vimeo.com/152631668. The performance documented there took place in Canada during the 2015 New Dance Horizons: Men in Dance Festival.

23 Zuzana M. Pick, *Constructing the Image of the Mexican Revolution: Cinema and the Archive* (Austin, TX: University of Texas, 2010), 202.

24 Lukas Avendaño, "Una aproximación a la muxeidad: queer: no. Queer-po muxe: sí," *Goethe-Institut: Mexico*, May 2019. See also, Lukas Avendaño, "Carta de un Indio Remiso," *Debates Indígenas*, 1 June 2022, https://www.debatesindigenas.org/notas/47-carta-de-un-indio-remiso.html.

25 For more about the current situation of Indigenous communities in Mexico, see the "2021 Análisis Situacional de los Derechos Humanos de los Pueblos y Comunidades Indígenas," *CNDH: Mexico National Commission for Human Rights*, "Activity Report 2021," http://informe.cndh.org.mx/menu.aspx?id=40067.

26 See Alejo Medina, "¿Dónde está Bruno?/Where is Bruno?" *Performance Research* 24, no. 7 (2019): 56–60; Antonio Prieto, "RepresentaXión de un muxe: la identidad performática de Lukas Avendaño," *Latin American Theatre Review* 48, no. 1 (2014): 31–53; Antonio Prieto, "The Queer/Muxe Performance of Disappearance: Lukas Avendaño's Butterfly Utopia," in *Performances That Change The Americas*, ed. Stuard A. Day (London: Routledge, 2021), 175–201; Maria Regina Firmino-Castillo, "The Refusal to be Disappeared: Lukas Avendaño's *Xibalbay*," *Howlround Theatre Commons*, 28 October 2021, https://howlround.com/refusal-be-disappeared-lukas-avendanos-xibalbay.

27 For more on the history of the Nahuas and its imperial advance through Mesoamerica, see Miguel Leon Portilla, *La Filosofía Náhuatl* (Mexico City: National Autonomous University of Mexico Press, 2016).

28 For more on the linguistic origins of the term "muxe" as well as greater detail about the gender dynamics and politics in Juchitan, see Marinella Miano Borusso, *Hombre, Mujer y Muxe en el Istmo de Tehuantepec* (Mexico City: CONACULTA-INAH, 2002) and Agueda Gómez Suárez and Marinella Miano Borusso, "Dimensiones Simbólicas sobre el Sistema Sexo/Genero entre los Indígenas Zapotecas del Istmo de Tehuantepec (Mexico)," *Gazeta de Antropología* 26, no. 23 (2006), http://www.ugr.es/~pwlac/G22_23Agueda_Gomez-Marinella_Miano.html.

29 For a more thorough discussion about the origins of the Zapotec language and the term "Binnizá," see Arthur A. Joyce, *Mixtecs, Zapotecs, and Chatinos: Ancient Peoples of Southern Mexico* (Oxford: Blackwell Publishing, 2009), 42–63.

30 Appler, "Goodbye Kiss."

31 See Blanca López de Mariscal, "La ejemplaridad en el auto de 'El Juicio Final.' Una lectura horizontica," in *Estudios sobre teatro español y novohispano de los Siglos de Oro: Actas del VIII Congreso de la Asociación Internacional de Teatro Español y Novohispano de los Siglos de Oro*, ed. Ysla Campbell (México: Universidad Autónoma de Ciudad Juárez, 1999), 83–92; and Diana Taylor, "Final Judgement," in *Stages of Conflict: A Critical Anthology of Latin American Theater and Performance,* eds. Diana Taylor and Sarah J. Townsend (Ann Arbor: University of Michigan Press, 2008), 48.

32 Felipe Cervera, "Theatre and Eschatological Politics," in *The Routledge Companion to Theatre and Politics*, eds. Helena Grehan and Peter Eckersall (London: Routledge, 2019), 295–8.

33 Eve Tuck and K. Wayne Yang, "Decolonization Is Not a Metaphor," *Decolonization: Indigeneity, Education & Society* 1, no. 1 (2012).

Creative Interlude

Mother

Chantal Bilodeau

An ageless woman

WOMAN

Look at me
Why won't you look at me?
You were such a surprise
No one was waiting for you
No one was expecting you
The idea of you
hadn't even crossed my mind
You just showed up one day out of nowhere
A fully formed creature
A little beast with its own will
ready to take the world by storm

I should have seen it coming of course
I should have realized when you crawled out of the ocean
that something was happening
that something had been set in motion

But the truth is
it was all random
A little mutation here
a little mutation there
and boom
there you were
with your big head and your big brain
so proud of yourself for standing on your own two feet

Such a miracle
Somehow
in the space of a geological flicker
chaos organized itself into you

Why won't you look at me?

Then one day
it was time for you to spread your wings
and make the world your own
And what a day that was!
Drunk with your own power
you heard and saw and tasted
You experienced and learned
Invented and discovered
Organized and catalogued
It was exhilarating!
I watched you grow
innocent and carefree
incapable of even imagining that
Well
maybe it wouldn't have made a difference

We used to gaze into each other's eyes
remember?
From dusk to dawn and dawn to dusk
I held you close and we lost ourselves into each other's mystery
That was before you started calling me Mother
Before there was anything to call Mother
I was you and you were me
There was no distance to travel
no otherness to name
It was our version of Eden
Our own little paradise
You know
Maybe the proverbial apple wasn't plucked from a tree
Maybe it was extracted from the ground
Look at me
I wish you would look at me
simply
without guilt or shame
I may be sick
I may be oozing thick black blood
but I'm not angry
I know you
For billions of years
I have carried you in my womb
You are made of me
You
my little piece of eternity
my little miracle

I know you're not ill-intentioned
I know you're not oblivious

I know you're not mean
You care
You want the best
You love
Yes you do
Maybe imperfectly
But you love

The past is behind us
Let it go
And the future
The future is forever hidden in the double helix of a cell
But now
Now is here
Now is us
Now is our time for forgiving what has been done
For rediscovering what has been lost
For healing what has been hurt
Now is our chance to bridge that distance again
and reclaim the Eden that was

before there was Mother
Look at me
Please
If I see myself in your eyes
I'll know that everything will be okay

PART THREE

Revising the Art-Science Repertoire

7

Performing and Negotiating Imperialism

Science, Agriculture, and Food in Puerto Rico

Teófilo Espada-Brignoni

Introduction

In April 1898, the United States declared war against Spain after the explosion of the USS *Maine* battleship, a long history of discussions and policy regarding expansion, and a deep-rooted desire to acquire Cuba.[1] American soldiers looked forward to fighting in the armed conflict, claiming they were in search of adventure.[2] A Kentucky newspaper even sent off one of its residents to Puerto Rico, celebrating how "the smell of [gun]powder is nothing new to Mike."[3] A couple of months later, the *Richwood Gazette* published a letter from an Ohio soldier fighting in Puerto Rico. After comparing the thickness of certain trees bearing familiar fruit to the hickory trees of Ohio, he adds, "They have a fruit here called mangos [*sic*]. They are about the size of oranges, but are a much finer fruit."[4] This quote preambles some of the ways in which Puerto Ricans were othered in similar terms to Edward Said's description of Orientalism, a colonial project grounded in an "ontological and epistemological distinction" between cultures.[5] The othering of countries, people, and food, particularly in the context of colonization efforts, serves to integrate the other as passive subjects whose existence is uncannily intelligible as an atavistic embodiment of personal, social, political, and technological life.

Between 1493 and 1898, Spain colonized Puerto Rico, decimated the native population, introduced slavery, and subjected the archipelago's economy to European markets.[6] By the 1890s, relations between Spain and the United States were tense and were exacerbated by the mysterious explosion of the *Maine*, which had sailed in Caribbean waters controlled, at the time, by Spain.[7] While there have been debates about what caused the explosion, many argued that Spain was not responsible for the tragedy, yet it provided the United States an excuse to declare war against Spain.[8] The war was officially over in December of 1898 after the Treaty of Paris. Afterward, the US Supreme Court declared that Puerto Rico was an unincorporated territory, meaning it belongs to the United States but is not part of the United States, articulating a legal framework for colonial relations.[9]

The Spanish-American War, and the subsequent acquisition of Puerto Rico by the United States as a territory in the Treaty of Paris, created an opportunity for complementary representations of Puerto Rico to develop in American and global fantasies of exploitation.[10] Even before the end of the war, US newspapers projected an increase in the market for publications about Puerto Rico due to the interest sparked by the conflict.[11] The writings that followed the war—some front stage, others backstage, and some in between, each with different target audiences—are interrelated textual performances that showcase the complex interplay among power relations, imperialisms, science, food, and accounts of Puerto Rico.

As Walter Mignolo argues, the Spanish-American War marks the beginning of the United States as an imperial power.[12] The war provided the opportunity for performances that sought to establish the United States as an empire and the positions and relations for other territories. As Said writes, "The main battle in imperialism is over land," and some of such battles were "reflected, contested, and even for a time decided in narrative."[13] The same can be said of official reports, correspondence, newspaper articles, and other documents, as these sources provide a window into the role of science in the articulation of the relationship between Puerto Rico and the United States and the discursive battles that ensued.

Work on the history of science in a colonial context, as Jorge Cañizares-Esguerra notes, has been limited.[14] In this chapter, I rely on notions about performance and the critical analysis of authors such as Cañizares-Esquerra, Mignolo, and Lisa Jackson-Schebetta, among others, to analyze the discussions surrounding the science of food and agriculture during and after the Spanish-American War. I look at the ways in which different sources (including books, newspaper articles, official reports, and letters) textually performed an image of Puerto Rico and Puerto Ricans that characterized US officials as benevolent actors and Puerto Ricans as passive individuals who should be grateful for the scientific and agricultural progress the United States promised the archipelago. As part of this study, I also pay attention to how imperialism was enacted through science and textual representations of Puerto Rico. Unlike other fruits, mangoes at the time were not well known in the United States, making them a mysterious object that required special attention (unlike well-known crops such as oranges or sugarcane that already existed in US farms and groceries).

Performance

The concept of performance in the social sciences and humanities is complex and multilayered, especially as it can simultaneously refer to several processes. In this chapter, my notion of performance draws from the work of Judith Butler and Erving Goffman, which I take as complementary accounts of how performances construct the world. According to Butler, performance is the "doing" that constitutes a social reality. As Butler argues, performativity relates not only to text but also to the body.[15] The same could be argued in this context, in which a performance includes not only the representation in print of the relationship between the metropolis and the other but also the ways in which imperialism is enacted through actual practices in the territories. By understanding agricultural science as performative, we can analyze its discourses and practices as "the spectacular and consequential ways in which reality is both reproduced and contested."[16] The colonized territory becomes the body of imperialistic performativity that is subjected to how metropolitan science constructs knowledge and becomes part of the infrastructure that allows one nation to exploit another one.

Following Goffman, performance is also that which is presented to observers in a social context with a specific purpose.[17] It is a version of the world that the performer expects to be taken by others as authentic. Through these processes, we see the ways in which "performative accomplishments take the place of nature."[18] Performances often (if not always) have not only a front stage but also a backstage. The front stage refers to the more or less ritualized and intentional (or unconscious) ways in which a performance is presented to others.[19] On the other hand, the backstage refers to the otherwise hidden aspects of a performance that play a role in setting up the front stage but are not meant for the audience.[20] We could consider published books, children's literature, or newspaper articles as the front stage and correspondence to and from officials of an agency as the backstage that played a role in designing and ritualizing the front stage.[21] At the same time, the backstage itself is another performance that, particularly in the case of private correspondence, relies on social conventions.[22] According to Butler and Goffman's theories, then, performances, which occur front stage and back stage, construct social reality in their own level of representation and inform each other while supporting or problematizing power relations.

In this chapter, I look at agricultural science in Puerto Rico from a performance framework. As performances do not occur in a vacuum, the documents associated with exporting mainland US science to Puerto Rico run across and parallel to other documents from other fields, including journalism and books for children. Thinking about science alongside other fields through a performance lens in no way disqualifies scientific practices. However, critical scholarly works in the fields of science and the humanities have underscored science *as* a culture and a site where performances about nature, human life, identity, authority, and society are negotiated.[23] Some modern Western scientists conduct their work partly under the assumption that an objective knowledge of nature's underlying laws would allow us to tame the world and likely improve it for human life.[24]

As Sandra Harding notes, traditional conceptions of science reject knowledges that fail to assimilate and conform to metropolitan practices.[25] Imperialism is grounded on the belief that colonial powers are superior and thereby have the right to subjugate others' knowledges and practices.

"The Better Order of Things"

As sociologist Lanny Thompson argues, the economic and political relations between the United States and Puerto Rico have been grounded in the ways in which Puerto Rico and Puerto Ricans have been represented in print media.[26] These representations mediated the relationships and perceptions between nations, institutions, and individuals. Science exists in and through this same network of relations: it sometimes supports or plays an active role in the domination of others,[27] and at other times, it resists harmful policies while still upholding the ontologies and epistemologies that helped to develop ecologically damaging industries in the first place.[28] Science can also be studied from Said's views on imperialism as "the practice, the theory, and the attitudes of a dominating metropolitan center ruling a distant territory."[29] From this perspective, some of the public, administrative, and private aspects of scientific endeavors parallel other institutions that integrate in their operations the representation of others.

US imperialism socialized audiences in North America and Puerto Rico into a system of relations in which the latter supposedly benefited from acting as obedient, colonized subjects. Journalism, children's literature, and science played corresponding roles in articulating such representations. In an 1898 newspaper article, writer and naturalist Frederick A. Ober described Puerto Rico as "a fat 'porker' with its legs chopped off."[30] This statement, if we consider the mapped silhouette of Puerto Rico as an inkblot that helps communicate the imperialistic mindset, embodies the imperial gaze toward the other. Puerto Rico was envisioned by Ober as a territory incapable of moving on its own and a place that allowed the observer to satisfy their greediest fantasies. A year later, Ober perhaps found the "porker" metaphor too colloquial or grotesque and instead wrote in his book that Puerto Rico "is almost a parallelogram in coastal outline."[31] Ober then intentionally changed a superficial aspect of his textual front-stage performance while supporting the same kind of political and economic domination. His writing embodied the textual performances through which Puerto Rico became a *body* (a butchered body, using his metaphor) to be dominated and consumed through the performative practices of imperialism. Certainly, his writing was not meant for Puerto Ricans, but it was part of the gradual ritualization of imperialism in print, as it was one of the front stages of US media.

Ober was only one of many who wrote about Puerto Rico. Another useful example comes from a 1900 children's book about Puerto Rico written by Marian George. In her book, George reproduced a front stage that invited US children to imagine potential roles for the Puerto Rican other and themselves. According to George, "the people [of Puerto Rico] seem very glad to take advantage of the better order of things."[32] George's book created two positions bonded by the assumption that one is the passive recipient of the supposed benefits of other (colonial) ways of organizing life. She expected colonized subjects to be grateful for how colonizers subjugate the other into their political and economic system. It is worth noting that George's idea of a "better order of things" resonates with the public (front stage) discourse of the officials associated with the creation of an Agricultural Experiment Station (AES) in Puerto Rico.

More than a stage, imperialism inserts itself and creates a multiplicity of auditoriums that inform each other and provide simultaneous versions of political relations to different audiences. Some of these auditoriums work as a front stage or backstage depending on the audience. These writings by Ober and George performed imperialism on different textual front stages (for adults and children) but can also be seen as the backstage that prepares metropolitan subjects for interactions with colonized individuals. Their views overlapped with official reports and correspondence among United States Department of Agriculture (USDA) officials, some of which constituted the backstage that laid the administrative and institutional infrastructure for the enactment of imperialism.

The Agricultural Experiment Station

While Ober's books were distributed, USDA officials hatched plans both front stage and backstage. In his 1899 annual report, the

Secretary of Agriculture suggested that AES should be established in Hawaii, the Philippines, and Puerto Rico.³³ AESs are fascinating institutions for they occupy and navigate an intermediary space between science, governmental interests, private business, and private citizens. Furthermore, as historian Richard A. Overfield argues in his study of the establishment of Hawaii's AES, many USDA officials adopted a model of agriculture that clashed with the preferred practices of the colonized territories.³⁴

In the United States, AESs were established through the Hatch Act of 1887.³⁵ According to historian David D. Danbom, AESs were supposed to provide valuable scientific advice to farmers based on their specific needs. AESs in the United States faced the challenge of "gaining rural acceptance for a standard of value which was essentially urban in nature,"³⁶ meaning AESs were viewed as outsiders by the communities they aimed to serve. Danbom's analysis resonates with that of Harding, who writes that sciences "have always been deeply integrated with their particular social and historical contexts. If they weren't they would be irrelevant."³⁷ Furthermore, AESs, in the context of imperialism, are an integral site for performances of colonization, turning the public aspects of science into a front stage for the performativity of imperialism. As Stuart McCook argues, after the Spanish-American War the United States used science "to consolidate effective control of the islands as an ideological justification for this control."³⁸

On June 13, 1900, the United States sent agricultural expert and educator Seaman Asahel Knapp to Puerto Rico on the US Army Transport *Burnside*, originally a Spanish vessel captured during the Spanish-American War.³⁹ Knapp was the special agent in charge of evaluating the possibilities of establishing an AES in the recently occupied territory. The US government sent the agriculturist as part of the continuing processes of transforming Puerto Rico's economy to serve the needs and desires of the North American markets. In 1901, the US government published and distributed Knapp's report.⁴⁰ At the time, Puerto Rican agriculture was in a precarious state due to several factors, including a hurricane in 1899 and the crippling debt to local, European, and US creditors that plagued most farmers.⁴¹ Knapp fails to mention, however, the many crimes committed against large plantations by gangs, some of which were originally organized by the US Army in an attempt to debilitate Spanish soldiers during the Spanish-American War.⁴²

Of course, if Puerto Rico had been perceived as a violent island, banks would not have assumed the risk to fund plantations as the performativity of imperialism implied the existence of obedient subjects.[43]

At the time of Knapp's report, there was not yet a university on the archipelago. For this reason, he suggested the sum of $15,000 in order to purchase land, erect buildings, and secure equipment, in addition to the $15,000 the Hatch Act could grant.[44] However, on March 2, 1901, the US Congress approved $12,000 "to establish and maintain an agricultural experiment station in Porto [sic] Rico, including the erection of buildings, the printing [. . .], illustration, and distribution of reports and bulletins, and all other expenses essential, to the maintenance of said station."[45] While in his report Knapp suggested the station should be built near the capital city, San Juan, it was ultimately established in Mayagüez in 1902.[46]

Knapp's report was far from an objective and uncontested description of the state and possibilities of Puerto Rican agriculture. He concluded by writing, "A few thousand dollars judiciously expended annually in the development of the agricultural resources of Porto Rico will result greatly to our benefit, because we shall then be buying our tropical imports with our surplus products."[47] Here Knapp outlined and described the agriculture and economy of Puerto Rico as something to be subjugated by the United States. Knapp believed Puerto Rico could be turned by the United States as a buyer for the mainland's surplus, which in turn would allow the United States to satisfy its desire for tropical products. The asymmetry of the relationship is quite clear: the metropolis enjoys desirable goods while the colonized territory, partly to meet the new demand, must adapt its infrastructure, agriculture, and scientific practices to meet the standards of the colonizer.

There is another detail missing in Knapp's report and other accounts of the state of Puerto Rican agriculture: the existence of the *Estación Agronómica* (Agronomic Station). Before the Spanish-American War, there were at least two stations in Puerto Rico, one in Mayagüez and one in Río Piedras. The local government ordered the construction of the *Estación Agronómica* in 1889, but construction of the station had already been approved since at least 1887.[48] Guillermo Quintanilla, a Puerto Rican engineer who studied agricultural science in Spain, was the director of the *Estación Agronómica* until 1897. Quintanilla was well respected in Puerto Rico and in Spain for his contributions to agricultural science.[49] The goal of the *Estación Agronómica*, as stated in official documents, was "researching scientific issues related to agricultural production in general, contributing to the scientific progress in speculative areas, and the divulgation through several means of the practical knowledge acquired" [my translation].[50]

The relative absence of information about this station in US official records is unsurprising, as marginalizing the contributions of locals would have reinforced imperialistic ideologies and justified textual and institutional performances of the station's importance. Furthermore, as McCook wrote, the United States' view of agricultural science conflicted with the research conducted in the Spanish Caribbean, where "the pursuit of scientific knowledge and the pursuit of practical goals" were supposed to be in harmony.[51] Ironically, before the Spanish-American War, agricultural research in Puerto Rico was inspired by the US model, which they adapted to the local context.[52] The mainland scientists, on the other hand, did not believe they had to adapt their knowledge and techniques. Latin American agricultural-scientific perspectives at the time, however, viewed the role of science as one of several approaches to dealing with "a wider array of agrarian problems, including shortages of labor," rather than producing knowledge in a vacuum.[53]

Imperialism not only constructs metropolitan subjects as the rightful captains, benefactors, and beneficiaries of the colonized; it constructs a place and specifies possible positions for the colonized. The question of the station and its funds also raised issues related to the politics of gratitude that, in the context of the performativity of imperialism, became the complementary set of representations that metropolitan subjects expected from the colonized. USDA officials saw themselves and the AES as key pillars in the improvement of Puerto Rico, realized through private and public textual performances. Both front and backstage, USDA officials expected Puerto Ricans to play a supportive role by showing their appreciation for what the station could do for them. US politicians wanted locals to show their appreciation for the station and to pay for it, despite having no say in how the station would operate. In a letter sent by A. C. True (director of the USDA Office of Experiment Stations) to Frank D. Gardner (who was appointed to direct the AES being established in Puerto Rico), True mentioned that the chairman of the House Committee on Agriculture said "that the people of

Porto Rico ought to do something to show their appreciation in this matter."[54] And in the station's first bulletin, Gardner wrote that for the AES to succeed, it needed "the interest, cooperation, and hearty support of the agricultural people of the island."[55] In private, USDA officials told Gardner how Puerto Ricans should behave regarding the AES; Gardner then relayed the message publicly to a broad Puerto Rican audience.

Negotiating Taste and How to Eat Mangoes

Imperialism perhaps influenced the eating habits of some US families, as the spaces of performativity grew and increased over time, reaching more people in the United States. According to historian Megan Elias, "The ingredients of American meals began to change significantly at the end of the 19th century as America expanded its territories around the world."[56] However, it is not merely that people in the United States suddenly became aware of produce from other lands. The USDA had an active role in laying the foundations for an import market in the United States, thereby creating demand for "exotic" tastes. The relationship between North American corporations, the Puerto Rican economy, US colonization, and changes to agriculture and food deserves to be explored in greater detail. As Jackson-Schebetta notes, "By 1922, 103 of the 107 major foreign corporations in Puerto Rico were US-owned, and the United States accounted for 90 percent of Puerto Rico's imports and 90 percent of its exports."[57] American demand for specific foods took precedent over the soil, population, farming, and scientific practices of other lands.

According to Knapp, most of the fruits he found during his travels in Puerto Rico were of great quality. "The mango is a luscious semiacid fruit, greatly prized in the Tropics," Knapp conceded; however, he continued, "It is valuable simply for home consumption."[58] It is worth noting that the mango is not originally from Puerto Rico but from India; they were gradually introduced to South America and the Caribbean during the colonization of the regions.[59] Being a tropical fruit, mangoes grow well and with little care in Puerto Rico.

Some of the properties of Puerto Rican mangoes were unappealing to US consumers, including its fibrous texture and the perceived messiness of the eating process.[60] Descriptions of the mango in the United States in 1898 suggested after eating one "it is necessary to take a bath and change the clothing in order to get rid of the superabundance of juice and pulp."[61] The taste and the fiber of local mangoes were not a source of contentious discourse in Puerto Rico, but the careless discard of the fruit's skin was discussed widely in the press due to the chance someone could slip and fall.

Before the Spanish-American War, a few scientists and travelers had some interest in the mango (and other "exotic" fruits). Botanist David Fairchild was one of the individuals responsible for the introduction of the mango to the United States.[62] Fairchild "sent USDA agricultural explorers, known as 'plant hunters' to collect thousands upon thousands of seeds and plants suitable for America's farms, home gardens, and city landscapes."[63] Horticulturist Elbridge Gale even grew mango trees during the 1890s in Florida; "Fairchild ate one of Gale's mangoes [. . .] It was so delicious that he vowed on the spot that he would persuade Americans that mangos were worth growing and eating."[64]

The station's views on mangoes did not change significantly over time. In 1905 the AES in Puerto Rico paid fifty cents for every thousand mango seeds Puerto Rican boys could provide for research; in total, they brought 10,000 seeds to the station.[65] However, in almost the same manner of Knapp's report, Henry Henricksen, the station's horticulturist, admitted that mangoes were not a priority and that his limited time was devoted to other experiments. One of the most interesting reports, however, is the one dedicated to mangoes in 1918. Charles Franklin Kinman begins this report as follows:

> The mango industry in Porto Rico has developed practically without attention or interest on the part of the inhabitants, although the fruit is one of the most important of the island [. . .] but all are of ordinary or poor flavor with an abundance of objectionable fiber in the flesh. With few exceptions the fruits are small, and this, together with their poor keeping qualities, makes them unsatisfactory as a commercial crop.[66]

After this, Kinman suggested that since most Puerto Ricans ate mangoes, it would be possible to create a market for this fruit within the island. This implied creating an unnecessary market, since as Kinman witnessed, and even as any traveler could also see today, there were innumerable mango trees in local roads in Puerto Rico. In 1903, the newspaper *La Correspondencia* reported that on the road from Ponce to Adjuntas, people could be seen picking up mangoes to satiate their hunger.[67]

The techno-agricultural and economic view of the mango also laid aside the cultural significance of the fruit in Puerto Rico. In 1903, the mango tree won a contest naming the most representative tree of the island. However, local schoolrooms and newspapers also asserted their own discourses and performances about Puerto Rico, which, in a way, served as physical and textual spaces for resistance. In an event called "La Fiesta del Árbol" ("The Celebration of the Tree"), schools celebrated the results of the contest.[68] Furthermore, the local version of the expression "low-hanging fruit" uses the word mango instead of a generic term for produce. In 1910, *La Correspondencia* published a short piece stating that mangoes were perhaps the most common fruit in tropical lands and that the increasing demands of English and US consumers were being met by India.[69] The article did mention that most varieties were not suitable for export, implying the reason was related to the damage the fruit could receive during transportation.[70]

Another act of resistance took place in the media after Knapp wrote in his report about sugar, a major export at the time. Local newspapers became a site for the resistance of North American performances in imperialism's front stages. According to Knapp, Puerto Rico lacked the infrastructure to produce enough sugar for the United States. However, Puerto Rican journalists and scientists were aware of Knapp's comments to American reporters and criticized his position in local newspapers. Some argued that Knapp's comments were due to a self-motivated desire to protect sugar plantations in Louisiana, in which he had interests.[71] In other words, Puerto Rican journalists problematized the authenticity of Knapp's account by inferring some of the potential backstage machinations and interests that motivated Knapp's discourse.

On Workforce

The performativity of imperialism also produces statements related to the body of the workforce tasked with satisfying imperial tastes.

The AES played a role in this context, mediating the relationship between North American diet and Puerto Rican agriculture by conducting experiments on what they could grow in Puerto Rico. But the experiments were not only to determine "many northern-grown vegetables are not adapted to tropical conditions."[72] AES officials also believed that it was important to conduct public experiments in Puerto Rico as a way of making Puerto Ricans passive recipients of scientific knowledge. Public experiments, according to Gardner, "create an interest in the immediate neighborhood where they are conducted, and furnish an object lesson for the people of the locality."[73] These experiments were among the performances meant to assert North American imperialism through science by evaluating Puerto Rican soil as a potential backyard for the United States. Furthermore, Puerto Rican workers were directed to produce specific manual performances that were meant to change their farming techniques. In his report, Knapp wrote that planters were not using appropriate scientific methods, which in turn negatively affected the amount and quality of the final crops or the products manufactured from them.[74] With more adequate methods, Puerto Rico would have had greater harvests and perhaps would have been able to meet the needs of the island as well as export goods to the United States. Concerns regarding the techniques used by Puerto Rican farmers (which would have been changed through public experiments) were, ultimately, about yielding enough produce to export to US stores and homes and meeting local needs.

In a letter to Gardner, True wrote, "In making plans for experimental work, you should consider especially the needs of the people of Porto Rico as regards the production of food supplies for home consumption, and the development of animal industry."[75] In this backstage communication, both seemed to believe in the beneficial role of an AES in Puerto Rico. At the same time, these considerations are not entirely innocent. Their analysis of how agricultural science could improve the lives of people in the archipelago was grounded in notions about the population and how to govern them, thereby positioning the United States as a benevolent empire. Following Goffman and Butler's ideas, I would argue that these dramatizations of benevolent imperialism are not always completely cynical. In other words, Gardner and his colleagues probably believed, albeit naively, that imperialistic performances and metropolitan assertions of power would better the lives of colonized people. Furthermore, regardless of their intentions and how they might have justified their actions, for themselves and others, they were an integral part of the turn-of-the-century imperial machine.

In his book, Ober wondered about the state of the Puerto Rican workforce, making remarks that served several functions at the same time: he constructed the United States as an imperial power deserving loyal subjects; criticized Spanish rule for how it sought to exploit Puerto Rico's resources; promoted the exploitation of Puerto Rican resources by the United States; and gave the appearance of an honest concern about the fate of Puerto Ricans if Spain had been able to fulfill its fantasies for the island. Ober's ambivalent views are evident in his description of how Spain treated Puerto Rico: "Doubtless, if the Spaniard had found all he expected here, the race inhabiting these islands to-day would be the most cruel on the face of the earth. [. . .] if these [exploitations] had continued, [. . .] the people of Puerto Rico would hardly be fit subjects for acquisition by the United States."[76] Here, Ober offers us a lens through which to read the other private and public documents in an imperialistic context. If officials had any actual concerns about locals, they were grounded in assumptions about the inferiority of the people they supposedly cared for but also wished to exploit. These otherwise sincere-yet-murky performances naturalized imperialistic discourses and practices and subjected the imagined well-being of the other to roles in service to the metropolis.

Inventing and Resisting Colonial Science

The Spanish-American War, the imperialistic attitudes it reinforced, and the efforts to colonize Puerto Rico sparked a multiplicity of interrelated discourses associated with diverse institutions. These discourses performed complementary accounts of Puerto Rico and gave meaning to the establishment of the AES and private investment in agriculture. Letters, official reports, and even books for children performed for different audiences how US officials and writers imagined and desired relationships between the United States and Puerto Rico in the years after the Spanish-American War. The United States textually invented Puerto Rico as a site of business and exploitation and Puerto Ricans as colonial subjects who should be grateful for what the United States decided for them. The way USDA officials talked about Puerto Rico in their reports and interviews for American newspapers made people want to invest in Puerto Rico and use science to profit from the new territory.[77] These performances of imperialism were enacted in multiple discourses and practices that amplified each other. However, local journalists and scientists problematized both the front and (inferred) backstage representations about the supposed supremacy of US-based science.

The experiments of the AES performed imperialism by inserting positivistic science in the lives and practices of rural Puerto Rico, reinforcing the notion that knowledge (and the practices that stem from it) is wielded and performed by the empire, not the colonized other. This created and reified, at the interpersonal level, ritualized politics, positions, and relations between the metropolis and the colonized territory. This is embodied in the desire to advertise the island's station as an institution that would provide answers to local problems while seeking to attract US investors. In doing so, the metropolis defended its textual and administrative performances of the social world as *the* one real and authentic performance.

Notes

1 Louis A. Pérez, Jr., *The War of 1898: The United States and Cuba in History and Historiography* (Chapel Hill: University of North Carolina Press, 1998), 5.

2 Harvey Rosenfeld, *Diary of a Dirty Little War: The Spanish-American War of 1898* (Westport, CT: Praeger, 2000), 29, 33.

3 "Mike at the Front," *The Central Record*, July 29, 1898, 1.

4 Artemus Sloop, "Another Version of the Battle of Guayama," *Richwood Gazette*, 25 August 1898, 2.

5 See Edward Said, *Orientalism* (New York: Pantheon Books, 1978), 2.

6 See Fernando Picó, *Historia General de Puerto Rico* (Río Piedras, PR: Ediciones Huracán, 1986).

7 Pérez, *The War of 1898*, 58.

8 Ibid., 63.

9 Rivera Ramos, "Deconstructing Colonialism: The 'Unincorporated Territory' as a Category of Domination," in *Foreign in a Domestic*

Sense: Puerto Rico, American Expansion, and the Constitution, ed. Christina Duffy Burnett and Burke Marshall (Durham: Duke University Press, 2001), 105.

10 Lanny Thompson, *Nuestra Isla y su Gente: La Construcción del "Otro" Puertorriqueño en "Our Islands and Their People"* (San Juan: Centro de Investigaciones Sociales, 2007).

11 Dexter Marshall, "Our New York Letter: A Great Revival in the Book Trade Is Expected This Fall," *Wilkes-Barre Times Leader*, 8 August 1898, 3.

12 Walter D. Mignolo, *Local Histories/Global Designs: Coloniality, Subaltern Knowledges, and Border Thinking* (Princeton: Princeton University Press, 2000), 136.

13 Edward W. Said, *Culture and Imperialism* (New York: Alfred Knopf, 1993), xiii.

14 See Jorge Cañizares-Esquerra, *Nature, Empire, and Nation: Explorations of the History of Science in the Iberian World* (Stanford: Stanford University Press, 2006), 7.

15 Judith Butler, *Undoing Gender* (New York: Routledge, 2015), 198–99.

16 Ibid., 30.

17 Erving Goffman, *The Presentation of Self in Everyday Life* (New York: Anchor Books).

18 Butler, *Undoing Gender*, 209.

19 See Goffman, *Everyday Life*, chapter 1.

20 Ibid., chapter 3.

21 Ibid., 22, 112.

22 Think, for example, of the behavior that would be considered appropriate or inappropriate in the dressing room of a concert hall.

23 Furthermore, many scientists have written critical accounts of science. See Steven Shapin, *Never Pure: Historical Studies of Science as if it was Produced by People with Bodies, Situated in Time, Space, Culture, and Society, Struggling for Credibility and Authority* (Baltimore: The Johns Hopkins University Press, 2010); Richard Lewontin, *The Triple Helix: Gene, Organism, and Environment* (Cambridge, MA: Harvard University Press, 2000); and Evelyn F. Keller, *Reflections on Gender and Science* (New Haven: Yale University Press, 1985).

24 Esther Díaz, "El conocimiento como tecnología de poder," in *La Posciencia: El Conocimiento Científico en las Postrimerías de la Modernidad.*, ed. Esther Díaz (Buenos Aires: Editorial Biblos, 2000), 19.

25 See Sandra Harding, *Objectivity and Diversity: Another Logic of Scientific Research* (Chicago: The University of Chicago Press, 2015), 22–3.
26 Thompson, *Nuestra Isla y su Gente*, 19.
27 See Harding, *Objectivity and Diversity*; Stephen J. Gould, *The Mismeasure of Man* (New York: Norton & Company, 1981); Todd May, *Between Genealogy and Epistemology: Psychology, Politics, and Knowledge in the Thought of Michel Foucault* (University Park, PA: Pennsylvania State University Press, 1993).
28 Consuelo Chapela, "From Latin America: About *Illusios*, the Rise of the Right, and Critical Qualitative Resistance," *International Review of Qualitative Research* 12, no. 1 (2019): 64, https://doi.org/10.1525/irqr.2019.12.1.60.
29 Said, *Culture and Imperialism*, 9; Laura Briggs, *Reproducing Empire: Race, Sex, Science, and U.S. Imperialism in Puerto Rico* (Berkeley: University of California Press, 2002).
30 Frederick A. Ober, "Crops of Porto Rico," *The St. Johnsbury Caledonian*, 28 September 1898, 4.
31 Frederick A. Ober, *Puerto Rico and its Resources* (New York: D. Appleton and Company, 1899), 11.
32 Marian M. George, *A Little Journey to Puerto Rico: For Intermediate and Upper Grades* (Chicago: A. Flanagan, 1900), 34.
33 "Secretary Wilson's Report," *The New York Times*, 30 November 1899.
34 Richard A. Overfield, "The Agricultural Experiment Station and Americanization: The Hawaiian Experience, 1900–1910," *Agricultural History* 60, no. 2 (1986): 258.
35 David B. Danbom, "The Agricultural Experiment Station and Professionalization: Scientists' Goals for Agriculture," *Agricultural History* 60, no. 2 (1986): 247.
36 Ibid.
37 Harding, *Objectivity and Diversity*, 2.
38 Stuart McCook, *States of Nature: Science, Agriculture, and Environment in the Spanish Caribbean, 1760–1940* (Austin: University of Texas Press, 2002), 48.
39 "American Money for Porto Rico," *The New York Times*, 14 June 1900; see also Wilhelm Hester, "Officers of USAT BURNSIDE in Port with Buildings in Background, Washington State, between 1899 and 1906," University of Washington Libraries, Special Collections,[HES007].

40 Seaman A. Knapp, *Agricultural Resources and Capabilities of Porto Rico* (Washington, DC Government Printing Office, 1901).
41 Ibid., 15.
42 Fernando Picó, *1898: La Guerra Después de la Guerra* (San Juan: Ediciones Huracán, 2013), 116. Writing such crimes in his reports could have had a negative effect on a decision to establish an AES, since many banks at the end of the nineteenth century sometimes refused to make loans to Puerto Rico, perceiving it as an unstable country.
43 Ibid., 33.
44 Knapp, *Agricultural Resources and Capabilities of Porto Rico*, 3.
45 US Congress, *The Statutes at Large of the United States of America, December, 1899, to March, 1901 and Recent Treaties, Conventions, Executive Proclamations, and the Concurrent Resolutions of the two Housses of Congress* (Washington, DC: Government Printing Office, 1901), 935–6.
46 United States Department of Agriculture, *Annual Report of the Office of Experiment Stations for the Year ended June 30, 1902* (Washington, DC: Government Printing Office, 1903), 163.
47 Knapp, *Agricultural Resources and Capabilities of Porto Rico*, 30.
48 "Construcción de Edificiones en Mayagüez Destinados á Instalar la Estación Agronómica," *Gazeta de Puerto-Rico*, 8 June 1889, 2.
49 María Teresa Cortés Zavala, "Agricultura Científica: Las Estaciones Agronómicas y la Caña de Azúcar en Puerto Rico, 1886-1897," *Ulua: Revista de Historia Sociedad y Cultura* no. 29, 73 (2017), https://doi.org/10.25009/urhas.v0i29.2538.
50 "Negociado–Fomento," *Gazeta de Puerto-Rico*, 18 September 1888, 1.
51 McCook, *States of Nature*, 3.
52 Ibid., 8–9.
53 Ibid., 3.
54 Correspondence from A. C. True to Frank Gardner, 6 March 1902. RG 164, Box 1, Folder Dept. Misc. Letters 1902, National Archives at New York City, New York.
55 Frank D. Gardner, *The Agricultural Experiment Station of Porto Rico; Its Establishment, Location, and Purpose* (Washington, DC: Government Printing Office, 1903), 14.
56 Megan J. Elias, *Food in the United States, 1890–1945* (Santa Barbara Greenwood Press, 2009), 11.

57 Lisa Jackson-Schebetta, *Traveler, There is No Road: Theatre, the Spanish Civil War, and the Decolonial Imagination in the Americas* (Iowa City The University of Iowa Press, 2017), 105.
58 Knapp, *Agricultural Resources and Capabilities of Porto Rico*, 22.
59 Deependra Yadav and SP Singh, "Mango: History and Distribution," *Journal of Pharmacognosy and Phytochemistry* 6, no. 6 (2017).
60 Ibid.
61 "Real Mango Trick," *The Buffalo Commercial*, 27 June 1898, 2.
62 See Robert R. Alvarez, "The March of Empires: Mangoes, Avocados, and the Politics of Transfer," *Gastronomica* 7, no. 3 (2007): 28, and Amanda Harris, *Fruits of Eden: David Fairchild and America's Plant Hunters* (Gainesville The University Press of Florida, 2015).
63 Alvarez, "The March of Empires," 28.
64 Harris, *Fruits of Eden*, 66.
65 H. C. Henricksen, "Report of the Horticulturist," in D. W. May, *Report of Agricultural Investigations in Porto Rico, 1905* (Washington, DC: Government Printing Office, 1906), 33.
66 C. F. Kinman, *The Mango in Porto Rico* (Washington, DC: Government Printing Office, 1918), 3.
67 "La Huelga de Panaderos de Ponce," *La Correspondencia*, 3 August 1903, 4.
68 El Corresponsal, "De San Lorenzo," *Boletín Mercantil de Puerto Rico*, 10 December 1903, 3.
69 "Lectura Campesina," *La Correspondencia*, 28 October 1910, 4.
70 Ibid.
71 E. Delafond, "Como se Escribe la Historia del Porvernir de la Industria Azucarera de Puerto-Rico en los Estados Unidos," *La Correspondencia de Puerto Rico*, 7 September 1900, 2. See also "Experimentos Agrícolas en Puerto-Rico," *La Correspondencia de Puerto Rico*, 17 September 1900, 2.
72 United States Department of Agriculture, *Annual Report*, 163.
73 Gardner, *Agricultural Experiment Station*, 12–13.
74 Knapp, *Agricultural Resources and Capabilities of Porto Rico*, 13–14.
75 Correspondence from A. C. True to Frank Gardner, 26 April 1901. RG 164, Records of the Office of Experiment Stations, Records Box 1, Folder: Misc Corres. 1901, National Archives at New York City, New York.

76 Ober, *Resources*, 164–5.
77 Correspondence from H. P. Happ to Seaman A. Knapp, 30 October 1901. RG 164, Box 1, Folder Dept. Misc. Letters 1901, National Archives at New York City, New York.

8

Identity Crisis in Interwar Germany

Brecht's *Leben des Galilei* and the Crisis of Science

Derek Gingrich

Bertolt Brecht's *Leben des Galilei* (*Life of Galileo*) is remembered as a meditation on the ethical responsibility of science in the shadow of the Second World War. But the play's life began before the Second World War as an exultation of the scientist as a folk hero.[1] Over eighteen years, Brecht revised *Galilei* numerous times, arriving at three distinct versions: the 1938 Danish version, written in exile from Nazi Germany; the 1947 American version; and the 1956 Berlin version, which he was revising when he died. English-language scholarship often treats *Galilei* as a quasi-autobiography, paralleling the development of Brecht's "*Theatre des wissenschaftlichen Zeitalters*" ("theatre for the scientific age"),[2] written and revised in light of Adolf Hitler's 1938 victories, the 1945 atomic bombings, and Josef Stalin's 1953 death.[3] English-language scholars often treat each edition as an iterative improvement, as when Cathy Turner appears to endorse Brecht's self-criticism of the Danish version before calling the final Berlin text more "richly intertwined" than prior versions.[4]

In this chapter, I argue that Brecht's 1938 *Galilei* goes beyond naively celebrating science. Rather, it critiques the anti-intellectualism endemic to German academia amid the country's interwar identity crisis. After Germany's defeat in the First World War, the educated middle class quieted their celebration of German science in a *Krisis der Wissenschaft* (crisis of science) that rattled the scientific establishment.[5] Public discourse embraced *Lebensphilosophie* (life philosophy), an anti-rationalist movement that encouraged Romantic notions of Germanness.[6] I contend that Brecht's Marxist and Taoist turns were an idiosyncratic response to the same identity crisis. As commentators pleaded for a revival of German Romanticism, Brecht championed the international proletariat as the site of crisis, rejecting an ethnic, geographic, and anti-Semitic definition of Germanness. In 1938, Brecht viewed physicists, especially atomic physicists, as kindred spirits. Scientific truth, according to his Marxist lens, revealed the contingency of mysticism, fascism, and theocratic power, thereby fueling revolutionary forces. The 1938 *Galilei* clarifies how Brecht understood the *Krisis der Wissenschaft*, outlining his own belief that mysticism must be fought by committing to the proletariat either actively or through principled Taoist nonresistance.

In this chapter, I explore the crisis of science in the Weimar Republic. Oswald Spengler's 1918 *Decline of the West* is emblematic of the rhetoric of *Lebensphilosophie*. Spengler argues that Western science had been forced to introduce theories (such as special relativity) that falsify the causal principles supposed by the Western scientific project, heralding the end of Western culture.[7] Chiding progressives like Marxists as delaying the inevitable, Spengler recommends superstition to soothe Europe's dying soul. Soon after, physicists (including Wilhelm Wein, Niels Bohr, and Werner Heisenberg), rebranded atomic physics as the science compatible with Spengler's era. As public commentators redefined German identity through historical poet-philosophers like Johann Wolfgang von Goethe (1749–1832) and Friedrich Nietzsche (1844–1900), physicists contextualized scientific discoveries while citing Goethe's work.[8] The language of *Lebensphilosophie* came naturally: prewar scientists routinely described science using Immanuel Kant's terminology, whose work influenced the scientifically minded artist of German Romanticism and offered an intellectual continuity between Wilhelmine and interwar ideas of German identity.[9] By joining this cultural movement, the physicists gave tacit support to an ethnically bound definition of German identity.

Second, I read the 1938 *Galileo* in the context of this identity crisis. In 1938, Brecht viewed atomic physicists as clever rogues pursuing scientific truth by going along with cultural currents, an idea he adapted from the Taoist principle of *Wu Wei*. The play lauds Galileo for outfoxing the Catholic Church by paying lip service to its theocratic ideals—a strategy Brecht commends because the revolutionary ends of science (as he understands it) justify the means of radical inaction. This understanding could not survive the atrocities of the Second World War. In subsequent versions of the play, Brecht unilaterally condemns the strategy. Thus, the 1938 *Galilei* stands as a cautionary tale about our capacity to justify inaction by focusing on the intended fruits of our intellectual labors.

Physics in the Hostile Intellectual Environment of the Weimar Republic

As Paul Forman outlines in his influential study of physics in the Weimar Republic, Germany's crushing defeat in the First World War instigated an identity crisis among its scientists, whose technical innovations seemed—when that war began—to assure victory. "Due to their contributions to Germany's military success," Forman writes, German physicists anticipated "a postwar political and intellectual environment highly favorable to the prosperity and progress of their disciplines" along with "public esteem and academic prestige."[10] At the precipice of the First World War, national identity and science were inextricable; scientists and humanists alike drew a line from Kant's rational philosophy to Germany's scientific output, advanced weaponry, and certain victory. A young Brecht joined the "general mood of patriotic fervour" sweeping the country as a pro-war poet steeped in scientific nationalism.[11] His first published poem, *The Kaiser/Silhouette*, weds German identity with Kantian rationality and technology. From the perspective of Germany's educated middle class, Kant set science, morality, and art on shared foundations, unified by human reason's ability to self-reflect. Thus, German physicists clarified the implications of new discoveries in Kantian terms.[12] In *The Kaiser*, Brecht demands that fellow Germans know themselves as heirs of Kant's legacy: "The Kaiser is Steep. Loyal. Steadfast. Proud. Just. / King of the land / of Immanuel Kant."[13] He extols "war, born from and giving birth to greatness,"[14] conflating forging a sword (an icon of technological superiority) with consecrating an altar. This pro-war rhetoric was commonplace: "Poets and artists everywhere joined the war effort," Stephen Parker explains, "articulating an idealistic belief in the necessity of a just war that would swiftly be won in the name of king and country, with God on their side."[15] Brecht saw Germany as the apex of European culture, honing itself through war.

But Brecht's faith in Germany's military waned as enlisted friends told him of the carnage at the front. His own short tenure as a medic expelled any lingering support. By 1917, Brecht's war poetry grew bitter, describing a soldier who "died with a groan and while not ready to die" as his beloved "searched on the dark roads and never found what she was looking for."[16] Through the early 1920s, Brecht underwent an identity crisis, rejecting the idea that Kantian lineage and martial technology defined the German spirit. Instead, he constructed new identities—German, artistic, and political—from eclectic sources. First, he turned to the German idealists, especially Friedrich Nietzsche and Arthur Schopenhauer, to reseat his idea of Germanness. He was particularly drawn to Nietzsche's *So Spoke Zarathustra* and its search for authenticity.[17] In Nietzsche, Brecht found something of a model: the individualistic genius who believes in radical transformation, self-possession, and integrity, rather than nationalism. His early works reflect Nietzsche's spirit, especially as dramatized in *Baal*'s iconoclastic hero (written 1918; premiered 1923).[18] Yet Nietzsche's lofty philosophizing left Brecht with a "hunger for the real,"[19] leaving him unsatisfied with German idealism beyond the self-possession at the heart of Nietzsche's works.

Second, Brecht sated that hunger and found a new political and artistic identity by engaging with Marxist thinkers, starting in 1920–1.[20] By 1928, he wrote about a "revolution in the theatre," promoting a "Proletariat" aesthetic of "class struggle" over the "bourgeoisie" aesthetic of traditional drama.[21] Understanding Marxism as fundamentally scientific, he opined that critics and theatre makers should "take a sociological and scientific standpoint" when thinking about theatre and its worldly effect.[22] Thus, Brecht arrived at an intellectual identity where the Romantic figure must leverage their genius to advance the proletariat cause—not abstract, aesthetic truths. Brecht's notion of authenticity expanded, too, beyond Nietzsche's pursuit of authentic selfhood to capture authentic, material conditions.

Third, Brecht followed Nietzsche's interest in Chinese philosophy, inuring him to ethnonationalist ideas of Germanness. As early as 1921, Brecht explored the Taoist principle of *Wu Wei* ("doing nothing") in his journal,[23] understanding the principle as "encountering fate with non-resistance":[24] an effortless action stemming from inaction. Brecht's Epic theatre, which intervenes in politics by accepting material reality on stage rather than resisting it, fuses this principled inaction with a Nietzschean quest for authenticity, that authenticity defined by a Marxist, proletariat lens. The trio shaped Brecht's understanding of German identity as a piece of a larger body (the proletariat), searching to understand its material conditions. This trio also shaped how Brecht identified the rebellious spirit—through *Wu Wei*, he embraced principled inaction, nonresistance, and passivity as possible routes to political action.

Brecht's Nietzschean flirtations mirrored the trend in the Weimar Republic. Germany's sudden defeat in the First World War saw public values "dramatically transformed."[25] Public figures openly linked the exact sciences (particularly physics) to Germany's *Entseelung*:[26] the destruction of its spirit. Soon, physicists from Max Planck to Max von Laue complained of hostility toward the sciences in private letters, public addresses, and open editorials.[27] Intellectual life exploded with talk of crisis. As Forman inventories, the *Krisis der Wissenschaft* became a "popular slogan" by 1921 in public discourse.[28] Thus, von Laue's 1922 *Deutsche Revue* editorial warned of pedagogues raising "serious accusations against modern natural science," as if science was "responsible for the crises facing the current world and spiritual and material misery related to those crises."[29]

Soon, a spate of intellectuals promoted *Lebensphilosophie* to rejuvenate German identity. A rejection of the Enlightenment more than a well-defined movement, *Lebensphilosophie* privileged vitalist categories like feeling, immediacy, and experience over analysis. As philosopher Jason Gaiger explains, *Lebensphilosophie* was indebted to Schopenhauer and Nietzsche, seizing Schopenhauer's vision of "life as all-encompassing metaphysical category" and Nietzsche's view that truth should be understood per "its function for life" not as a thing awaiting discovery.[30] As a Marxist, Brecht could still align with the late-career Nietzsche of *Nietzsche contra Wagner* (1889), who condemned the nationalistic underpinnings of his own early work.[31] But the Weimar (and Nazi) public embraced the Nietzsche of the *Birth of Tragedy* (1872), who lauds the anti-Semite Richard Wagner as the savior of the arts: a troubling endorsement when read across from Nietzsche's *Genealogy of Morality* (1887), which permits ethnonationalist interpretation.[32] In other words, pulling from the same inspiration as Brecht, a spate of the public instead developed an increasingly ethnonationalist idea of Germanness.

No figure spread the sentiments of *Lebensphilosophie* further than Spengler, whose *The Decline* declares Western science dead. *The Decline* forwards that cultures are "organic forms" with natural life cycles: each culture emerges, navigates its ineffable existence, dies, and is then replaced.[33] Moreover, a culture brings about "its own possibilities of expression, which appear, ripen, decay, and never return"—these expressions capture a culture's true identity, not any sort of reality they try to describe.[34] A culture expires after it fully maps the contours of its soul through its mode of expression. Spengler cites two sources for his theory: Goethe, who taught him a distinct scientific method; and Nietzsche, who taught him that culture is an integrated system with slaves and masters.[35] With Nietzschean pessimism, Spengler sees the end of Western culture coming. As Spengler defines it, Western culture began in tenth-century Europe and (by 1918) almost encompassed the planet. Its "Faustian" spirit is crystallized in the principle of causality, which reduces everything to material cause and effect. Math and physics were the West's mode of expression, and he chides scientists and mathematicians as "abstract academics" whose "entire existence rests on the principle of causality."[36] Spengler argues that the end of Western science is implied by contradictions brought about by thermodynamics, atomic decay, and Einstein's theory of relativity.[37] These advances revealed the "concepts of mass, space, absolute time, and natural laws in their entirety" as expressions of culture, not facts.[38] Spengler expects Westerners to embrace a "second religiosity," warming their "geriatric souls" as their culture dies.[39]

The influence of Spengler's book cannot be understated. In five years, the first volume sold "100,000 copies in a country with scarcely three times that number of college graduates."[40] "If the scholar or scientist was to maintain his prestige," Forman explains, he had to "repudiate the traditional methods and doctrines of his discipline."[41] Physicists denounced "positivist conceptions of the nature of science, a utilitarian justification of the pursuit of science, and, in some cases, the very possibility and value of the scientific enterprise."[42] Even conservative materialists like Wein leavened scientific findings with *Sturm und Drang* in public lectures.[43] By 1929, Hans Geiger et al.'s *Handbook of Physics* called science an irreducible and unanalyzable human drive,[44] presenting scientific inquiry in the language of *Lebensphilosophie*. By 1930, Heisenberg's description of quantum mechanics recommended abandoning "our ordinary description of nature, in particular the idea that processes of nature are strictly causal," an invective supported by his then-mentors Max Born and Bohr.[45] Some physicists rejected Heisenberg's *Lebensphilosophie*-inflected description of atomic physics—Einstein and Erwin Schrödinger most vocally.[46] However, even after relations between Heisenberg, Bohr, and Born cooled due to the former's compliance with the Nazis during the Second World War,[47] the elder physicists still described physics in Heisenberg's language.

The physicists did not invent quantum mechanics because of Spengler. Rather, *The Decline*'s popularity generated an atmosphere among German intellectuals, dubious of fixity, in which quantum mechanics thrived because it offered cultural relevance when other exact sciences did not. In public, physicists emphasized the unpredictability of quantum events and linked that unpredictability to free will, vitalism, and human imagination in a project of *Lebensphilosophie*.[48] This rhetoric glosses over quantum mechanics' linear evolution of probabilities: a deterministic process contra Spengler's prognosis. In scientific publications, physicists still used causal language—for example, in 1930, Heisenberg suggested interference from the measuring apparatus *caused* quantum acausality.[49] As Germans responded to the postwar identity crisis by retreating to the Romanticism that fueled prior German nation-building projects, atomic physicists followed suit to maintain cultural capital. But they did not seriously disrupt their work. Nor did this shift require significant intellectual labor: after all, German Romanticism was indebted to Kant's theory of imagination. For many physicists, the change simply required them to replace

Kant's language of critical reason with his aesthetic judgment and imagination, as presented by Goethe. Brecht started engaging with science amid these upheavals. Because he saw Marxism as a scientific, rational enterprise of social justice, he rejected Spengler's anti-scientific proclamations. He may laud "Spengler's great book"[50] in 1920 at the height of his Nietzschean phase, but by 1926, Brecht's *More Good Sports* repudiates Spengler outright.[51] *Lebensphilosophie*'s focus on the individual, appetites, and lived experience reflected Brecht's own inclination, but Brecht needed the philosophy's tenets to serve a Marxist revolution by encouraging workers to identify as a unit with similar biological drives. Moreover, Spengler was staunchly anti-Marxist, anti-Semitic, and a prominent member of the Conservative Revolution: an anti-democratic movement promoting a new nationalism based on the German state as a unified organism, formed by the land, whose natural order was dictatorship and ethos Nietzschean. Similarly, Heisenberg led a Path Finder youth group, which indoctrinated young boys with nationalistic values. Neither man joined the Nazi Party: Spengler resigned from his academic position to protest the Nazis' racialist policies, despite his own anti-Semitic beliefs;[52] Heisenberg declined party membership despite overseeing a fission research unit under Nazi supervision and advising party officials on the topic.[53] During the war, he told physicist Hendrik B. G. Casimir that German victory was "perhaps" the "lesser evil" of the war's foreseeable outcomes: Europe run by Hitler or Stalin.[54] Yet he rebuked Nazi physicists Philip Lenard and Johannes Stark's "Aryan physics" program—the dismissal of Einstein's theories as "Jewish constructs," not science.[55] Because Heisenberg continued teaching Einstein, Stark condemned quantum mechanics as "Jewish physics," branding Heisenberg a "White Jew" who deserved prosecution.[56] Regardless, Stark and Heisenberg both directly benefited the Third Reich.

When Brecht wrote the first *Galilei* in 1938, he recognized the protofascist leanings of Spengler's ilk, but he assumed that scientists were aligned with the proletariat spirit, like himself. Emerging from the *Krisis der Wissenschaft*, *Galilei* justifies their acquiescence toward mystical language (first) and eventually Nazism (second) to dissolve the contradiction between their behavior and the Marxist nature of their project. *Wu Wei* offered Brecht a way to justify this strategy: their nonresistance permitted indirect action via scientific discovery.

The Danish *Life of Galileo* and Quantum Mechanics

Brecht's presentation of Galileo in 1938 reflects his specific citation of physicists in the interwar period. He first demonstrated familiarity with quantum mechanics in 1927, when he declared that his play *Mann ist Mann* (*Human Equals Human*) explored how "the continuous self is a myth. A person is an atom, always decaying and forming anew."[57] Before the rise of the Nazi Party, Brecht's understanding of atomic physics echoed Spengler's language: characters decay like cultures, a new character taking their place. These atomic proclamations come only a year after Heisenberg published for the first time on transition probabilities and quantum jump,[58] concepts that Brecht soon mobilized aesthetically.[59] Brecht embraced Heisenberg's famous uncertainty principle, as explained in the *Principles of Atomic Physics*. According to Heisenberg, some properties of particles possess a limited maximum accuracy: most famously, position and momentum. Because the scale of atomic phenomena is so miniscule, there remains a fundamental uncertainty regarding an atom's position or momentum even at the theoretical maximum precision. In *Principles*, Heisenberg hypothesizes that uncertainty arises from the interaction between the measuring instrument and the atomic object, which acts upon (and therefore changes) the atom, adding a fundamental unpredictability to the results.[60]

Brecht reveals the importance of atomic physics to his thought in *Flüchtlingsgespräche* (*Refugee Conversations* (written circa 1940)), an unpublished dialogue. Ziffel, a physicist character and Brecht's stand-in, explains Heisenberg's uncertainty principle:

> The light in the microscopes is so strong that it causes revolutionary heat and chaos. . . . We do not see the normal life of the atomic world, but its life disturbed by our observation. . . . Similarly, social science does not leave the processes of the social sphere untouched: it has a strong effect

on them. A revolutionary effect. Thus, authoritative circles reject social science.⁶¹

Brecht extrapolates from Heisenberg's causal explanation of uncertainty that if studying an atom *causes* revolutionary change, studying people similarly instigates revolution. This revolutionary reading of quantum mechanics encouraged him to view the physicists themselves as revolutionaries.

Heisenberg's textbook presents uncertainty alongside Bohr's complementarity theory: the idea that there are multiple, equally correct ways to present quantum systems. One might describe atoms in space and time, necessitating uncertainty because the experimental apparatus changes the system being studied; or one might express phenomena under the rubric of causal laws, which abstracts them from space, time, and the experimental condition.⁶² Brecht's Epic theatre echoes these modes: the dramatic experiment lets audiences study the play as directly as possible but eschews a certain or fixed narrative voice, while Epic interventions leave the fictional space and enter the realm of abstract commentary.

The Danish *Galilei* exclusively occupies the empirical mode of space-time and uncertainty, lacking the Epic interjections that characterize the later editions and clarify Brecht's Marxist lessons. Three features primarily differentiate the Danish version from its successors. First, there are no singers or placards announcing subscenes, which push spectators into foregrounding the explanatory, abstract mode. The play only shows the experiment of drama, avoiding, compared to Brecht's other works, theorizing. Second, it lacks the explicit critique of capitalism that motivates later editions, focusing on the threat of theology. Third, the Danish Galileo figure is a magnetic force, attracting audiences with charm and wit as he evades authority and champions truth, whether in public debate or under the Inquisition's thumb. As the *Handbook of Physics* puts it, this Galileo feels an irreducible drive, a veritable hunger, for knowledge and earthly delights comingled.⁶³

The Danish *Galilei* opens on Galileo washing himself, as Andrea, his housekeeper's son, enters with a glass of milk and a wake-up call. Galileo, "snorting and joyous," savors his milk like an Epicurean gourmand, indulging his body, then eager to indulge his mind with a book.⁶⁴ Andrea wants to pay the milkman, but Galileo has a present for the boy: a model of the Ptolemaic system, in which the planets, sun, and stars revolve around the earth, embedded in crystalline spheres. They launch into Socratic dialogue, Galileo guiding Andrea toward a realization of the model's flaws. He illustrates his alternative theory by placing Andrea in a chair to physically drag him around the sun. For a moment, *die Erde bewegen sich* ("the Earth moves"), the play's original title, from Galileo's own strength and individual verve.

Right away, Galileo combines *Lebensphilosophie*'s vitality, Nietzsche's iconoclastic individualism, and a Marxist's revolutionary urge to educate the proletariat. Moreover, he functions as the atomic scientist-hero, echoing Brecht's understanding of quantum mechanics. Christian Møller (Bohr's lab assistant in Copenhagen) helped Brecht understand both the science covered by his *Galilei* (i.e., Copernican astronomy) and, subsequently, atomic physics during Brecht's time in exile.[65] The play's presentation of Copernican physics overlays the two. Placing Andrea into the center of the experiment, sat upon the "Earth," Galileo enacts Bohr's declaration that "any sharp separation between the behaviour of atomic objects and the interaction with the measuring instruments which serve to define the conditions under which the phenomena appear."[66] Galileo actively connects his observations to the needs of the commons: he *chooses* to share his revolutionary ideas with his servant's son, on the principle that the boy deserves better than ignorance.

In comparison, singers introduce the 1947 American *Galilei*, deflating Galileo's lesson by summarizing it beforehand. Brecht maintains a critical distance, like quantum mechanics' causal mode of explanation, securing moral certainty at the cost of engaging with Galileo's gravity as a personality. Andrea enters with the astronomy model, a gift from the Neapolitan court, and a colder Galileo dismisses such "stupid presents," wishing they had sent money.[67] Then, Andrea, the working-class boy, tricks a mercenary Galileo into teaching him by demonstrating his eagerness. When Galileo's second student arrives for a lesson, each version of *Galilei* produces a student indicative of its unique preoccupations. In the Danish *Galilei*, Doppone wants to become a theologian, a career pushed on him by his father because Doppone "likes to argue."[68] Doppone's theocratic contrarianism irks Galileo, foreshadowing the theocracy that derails Galileo and (by metaphor) modern physics. The American *Galilei* features Ludovico, a merchant's son whose mother has sent him to learn science to be more pleasant at parties. Galileo recognizes Ludovico as a source of income. The Danish *Galilei* draws similarities between the mysticism physics faced in Weimar Germany and the religiosity faced by Galileo; Doppone is no callous agent of the state but a young boy swept up in anti-scientific currents. Rather than resisting, however, Galileo adapts; he teaches the boy and takes his money to fund his scientific work. This Galileo reflects Brecht's assessment of atomic physicists, adapting to dogmatic environments as they arise to continue their work in peace.

Theocratic fascism demands that one ignore all evidence not supporting existing orthodoxy. Arriving in Florence, the Danish Galileo is accosted by a mathematician, a philosopher, and a theologian who demand logical explanations for Galileo's heliocentric theory. He promises proof if they look through his telescope. The mathematician, tempted, shies away, afraid of the ramifications. The philosopher refuses because he loves the simplicity of Aristotle's universe (which places him at its center). The theologian "pulls out his handkerchief and, with a significant look at Galileo, washes the lens,"[69] leaving with a smirking implication that either Galileo's instrument was dirty or Galileo painted his "proof" onto the lens. A man approaches an experimental tool, reminding spectators that he *could* look—Ziffel's revolutionary measurement, tapping the generative power of uncertainty. But he refuses, rejects science, protects Ptolemaic tradition, and props up authority. The *gestus* demonstrates the theocratic ability to walk to the brink of scientific knowledge, be it physics or Marxism, then publicly shame it as felonious without an honest peek.

Identifying these gestic characters helps determine the scope of Brecht's criticism. Quantum mechanics, too, faced institutional resistance, first from other physicists and then from the Third Reich. From 1927 to 1935, Einstein and Bohr debated the young theory, with Einstein leading the contingent rejecting it in the language of *Galilei*'s mathematicians. Einstein demanded causes, insisting that "our dear God does not play dice."[70] When *Galilei*'s mathematician demands a causal account, rather than empirical observations, the play recalls the Bohr-Einstein debates, likely explained to Brecht by his Danish hosts. German literature scholar Erhard Bahr argues that the "model for Brecht's Galileo was the Danish physicist Niels Bohr."[71] But Brecht's allegiances here are not so cut-and-dried. Brecht sympathized with Einstein's insistence on causality because of their shared commitment to socialism. In *Me-Ti: Buch der Wendungen* (*Book of Twists and Turns*), a posthumously published collection of Taoist aphorisms Brecht began in the late 1930s, he warns physicists to hold tight onto the principle of causality. As a Marxist, he recommends studying people through the lens of causality, not analyzing nature with the concept of free will.[72] By 1941, he bemoaned the mystical tendencies in recent atomic physics publications—Romantic thinking was not compatible with revolutionary spirit.[73] Finally, the Danish *Galilei*'s gradation of institutional resistance (unique to this version) offers a more nuanced story. The philosopher resists because he wants to be the center of the universe *as such* (not merely the center of measurement). Galileo's evidence demonstrates his mundanity. He is the mystical thinker, looking to astrology rather than astronomy. The theologian resists because empirical evidence threatens his power—he understands that mysticism is something to be weaponized. But the mathematician *almost* looks: his *gestus* of approach suggests a scientific mindset trapped by self-preserving cowardice, not malice or opposition. He knows he cannot publicly accept Galileo's findings without danger. He could pursue truth, if incentivized. In contrast, the American version's mathematician denounces Galileo as a fraud to keep his court pension. By 1956, Brecht had decided that a weakness for bourgeois comfort, not self-preservation, guided the scientists' actions.

Galileo's findings are both confirmed by the Inquisition and condemned by it, a quantum state in which Galileo dwells for the play's midpoint. Like the physicists in the Weimar Republic, Galileo continues his research, with full support of the Vatican, but he may not publish anything that contradicts dogma. He continues his revolutionary science in private, and he tailors his public image to fit the cultural directive. However, the next pope pushes the Inquisition to force Galileo's recantation; he capitulates immediately. The Weimar Republic ends, and the Third Reich begins. After recanting, Galileo explains his decision by telling his students a parable cut from later editions. After a tyrant takes Crete, his agent enters Keunos's home with a writ stating that "any home he set foot in belonged to him, as did the food, and the people there must serve him."[74] He asks Keunos to serve, and the philosopher wordlessly complies for seven years, until the agent dies from "excess food, sleep, and demands."[75] Keunos rolls the body in a dirty sheet, cleans up, and finally answers: No. As Parker highlights, the Keunos myth recapitulates Brecht's understanding of *Wu Wei*, the action through nonresistance central to Taoism.[76] In the *Wendungen*, Brecht places *Wu Wei* alongside quantum mechanics and Marxism: for Brecht, quantum mechanics teaches us that scientific study is a kind of effortless action, a subtle but revolutionary intervention.[77] Galileo's resistance is Taoist, atomic, and revolutionary despite capitulating.

Brecht's Danish *Galilei* dissolves the contradiction between his belief in science's inherent Marxism and the atomic scientists' capitulation to power: physicists are stealthy iconoclasts, atoms whose revolution comes through observation and study. Their nonresistance surfaces the truths about power that fuel action. When *Galilei* shows a change in popes, Brecht implies that the atomic physicists remaining in Germany are continuing the strategy they honed during the Weimar Republic despite changing circumstances—from the anti-intellectualism of *Lebensphilosophie* to the horrors of Nazi rule. In *Fear and Misery of the Third Reich* (premiered in 1938), Brecht explicitly dramatizes this dynamic. Two German physicists receive correspondence from Einstein. With his notes, they excitedly advance a theory of gravitational waves, though the pair is constantly interrupted by the need to listen intently for surveillance, "examining the telephone, checking out the door, etc."[78] After one accidentally names Einstein, the other loudly and performatively denounces Einstein's "Jewish nit-picking" before the pair continue quietly.[79] Their acquiescence, Brecht suggests, is a show.

Brecht likely had Heisenberg in mind when he wrote *Fear and Misery*—Stark condemned the physicist as a "White Jew" less than a year before the play's premiere.[80] Of course, Brecht could not know the label's toothlessness—Heisenberg's possible usefulness to the Nazis outweighed Stark's condemnation. Postwar, many physicists presented themselves in a similar manner to *Fear and Misery*. As Stephan Schwarz outlines in "Defending Alignment," physicists like Heisenberg and von Laue defended their decision to remain in Germany as an effort to save science, claiming they minimally complied with Nazi orders to show "political reliability" and avoid persecution.[81] But Schwarz exposes the physicists' justifications as convenient rationalizations, morphing to fit the context. Heisenberg started with "non-committal wishful thinking," writing in 1933 that the "ugly element" (i.e. anti-Semitism) of Nazism was a passing phase, to historiographical justifications in 1942, false dichotomies (party vs. opposition as the only options) during the de-Nazification process, and vague metaphors likening physicists to ships "forced out into a hurricane" without reference to material reality in 1969.[82] Schwarz labels Heisenberg's rationalization as, at best, "pragmatic survival strategies" rather than secretive resistance.[83] Far from *Wu Wei*, the physicists sought to exonerate themselves, their true motives a complex web of familial concern, fear, national pride, arrogance, and prejudice.

Heisenberg's denouncement never seriously endangered him, but Brecht emphatically identified with the plight of the Jews, both in his personal life through his wife Helene Weigel (who was Jewish) and in plays like *Fear and Misery*. Also, he actively resisted the Nazi regime. Unlike Galileo and Heisenberg, he fled, producing public, anti-Nazi art. The final scene of the Danish *Galilei* makes space for both strategies. Andrea smuggles Galileo's *Discorsi* across the Italian border. There, local boys argue about a witch who lives in a nearby house, claiming they have seen her fly away on a broomstick and that she never leaves her home for food (contradictory beliefs, of course). Their prejudice stems from misunderstanding and misinformation—an analogue to anti-Semitic beliefs based on misinformation and conspiracy. Andrea shows the boys their error, revealing a hidden delivery jug on the porch and proving she has food delivered. She eats. With the boys, he imagines a bright future where science erases false beliefs: "we stand at the beginning"[84] of a scientific utopia. To the revolution, Galileo contributes his intellect: he thinks, studies, and discovers the truth. And Andrea spreads the truth, as Brecht and his troupe revealed the lived experience of Jews under Nazi rule through their theatre. Some must stay; others must flee. In the Danish *Galilei*, the answer to German identity crisis is in revolutionary spirit, inherent in scientific ways of knowing, whether that science is theoretical (like physics) or practical (like theatre for the scientific age).

Conclusion

After the atomic bombs dropped and the horrors of the Holocaust were made plain, Brecht removed this final scene. When the Danish Galileo first recants, Andrea laments, "Unhappy is the land with no heroes!"⁸⁵ Galileo playfully retorts: "No. Unhappy is the land that needs heroes."⁸⁶ In 1938, Brecht tells us we may not recognize heroes—the land is unhappy. Individuals adapt, survive, and rebel in idiosyncratic ways. The American Galileo lacks this impish spirit. He is an obsessive, self-absorbed scoundrel wracked with regret. The American Andrea forgives Galileo because (he explains) productive hands stained with complicity are better than empty hands that have produced nothing for the revolution.⁸⁷ But Brecht and Galileo refuse forgiveness. Galileo castigates his own inaction, his decision to study instead of defending rationality in the streets. He laments the "new ethics" of "common sense" survival brought with his "new science."⁸⁸ Brecht leaves his audience reflecting on Galileo's complicity developing new weapons. The uncertainty principle now carries ethical weight. If choosing to study something is a radical social act, then how do we interpret the American-aligned scientists, who developed the atomic bomb? How do we understand the Nazi-aligned scientists, who helped the Nazi war effort despite the atrocities of the holocaust? The audience is left uncertain, bereft of Andrea's revolutionary praxis at the border crossing.

But the Berlin version revives the border crossing.⁸⁹ Andrea lacks the pluck of the Danish version, but he still spreads the *Discorsi*. Identity is not enough, Brecht decided. Action is necessary. In 1938, Brecht excused scientists' inaction because he saw their identity as inherently revolutionary. He conflated scientific progress with Marxism as he drafted his theatre for a scientific age and smuggled it across the border. Rewriting *Galilei* over the next eighteen years, Brecht identified himself and others through their actions. Is Galileo a revolutionary if he failed to defend rationality in the streets? Or was his private revolution an excuse for cowardice? Pursuing the truth is not enough. We must shout it in public and throw it in the face of power.

Notes

1 Bertolt Brecht, *Leben des Galilei: drei Fassungen, Modelle, Anmerkungen* (1943; Berlin: Suhrkamp, 1998).

2 Bertolt Brecht, "Kleines Organon für das Theater," in *Werke. Große Kommentierte Berliner Frankfurter Ausgabe* (1949; Berlin: Suhrkamp,

1998), 23:65. Subsequent German-language citations appear in my translation.

3 John Willet, "Introduction," in *Life of Galileo*, by Bertolt Brecht, eds. John Willet and Ralph Manheim, trans. John Willet (New York: Penguin, 2008), xli.

4 Cathy Turner, "Life of Galileo: Between Contemplation and the Command to Participate," in *The Cambridge Companion to Brecht*, eds. Peter Thomson and Glendyr Sacks, 2nd ed. (Cambridge: Cambridge University Press, 2009), 146, 148; see also Willet, "Introduction," xliv–xlv.

5 Paul Forman, "Weimar Culture, Causality, and Quantum Theory, 1918-1927: Adaptation by German Physicists and Mathematicians to a Hostile Intellectual Environment," *Historical Studies in the Physical Sciences* 3 (1971): 8–9.

6 Ibid., 26–8.

7 Oswald Spengler, *Der Untergang Des Abendlandes*, 1st ed. (München: C.H. Beck'sche, 1919), 29.

8 For example, Max Born, *Die Relativitätstheorie Einsteins und ihren physikalischen Grundlagen*, 1st ed. (Berlin: Springer, 1920).

9 Kant's description of cognition reconciled abstract mathematics with the empirical aims of natural science. For example, Gustav Doetsch, "Sinn der reinen Mathematik und ihrer Anwendungen," *Kant Studien* 29 (1924): 439–59.

10 Forman, "Weimar Culture," 8–9.

11 Stephen Parker, *Bertolt Brecht: A Literary Life* (London: Bloomsbury, 2014), 57.

12 For example, Max Planck began his Columbia lectures with a Kantian discussion on a priori laws. Max Planck, *Eight Lectures on Theoretical Physics*, trans. A. P. Willis (New York: Columbia University Press, 1915), 11–13.

13 Brecht, *Werke*, 13:76.

14 Ibid.

15 Parker, *Bertolt Brecht*, 58.

16 Brecht, *Werke*, 13:87.

17 Ibid., 26:108.

18 *Baal* depicts an amoral poet who revels in unrestrained freedom, even as that freedom kills lovers, friends, and eventually himself: the inherently tragic life found in Nietzsche's work.

19 Brecht, *Werke*, 26:108.

20 Ibid., 26:114.
21 Ibid., 21:233–4.
22 Ibid., 21:232.
23 Ibid, 26:262.
24 Ibid., 26:580.
25 Forman, "Weimar Culture," 9.
26 Ibid., 10.
27 Ibid., 12–19.
28 Ibid., 26–8.
29 Max von Laue, "Steiner und die Naturwissenschaft," *Deutsche Revue* 46 (1922): 41.
30 Jason Gaiger, "Lebensphilosophie," in the *Routledge Encyclopedia of Philosophy*, ed. Craig Edward (New York: Routledge, 1998).
31 Friedrich Nietzsche, *Nietzsche contra Wagner: Aktenstücke eines Psychologen* (Leipzig: C.G. Neumann, 1889).
32 Nietzsche later wrote a forward condemning Wagner, which was added to a posthumously published edition. Friedrich Nietzsche, "Vorwort an Richard Wagner," in *Die Geburt die Tragödie* (Berlin: Walter de Gruyter, 1967).
33 Spengler, *Untergang*, 1st ed., 29.
34 Ibid.
35 Oswald Spengler, *Der Untergang Des Abendlandes*, 2nd ed. (München: C.H. Beck'sche, 1923), xiv.
36 Spengler, *Untergang*, 1st ed., 168.
37 Ibid., 596–7.
38 Ibid., 597.
39 Ibid., 609.
40 Forman, "Weimar Culture," 30.
41 Ibid., 28.
42 Ibid., 7.
43 Ibid., 10–11.
44 Hans Geiger, Karl Freidrich Franz, and Christian Scheel, *Handbuch der Physik* (Berlin: Julius Springer, 1928), 1–2.
45 Werner Heisenberg, *Die physikalischen Prinzipien der Quantentheorie* (Leipzig: Von S. Hirzel, 1930), 48.

46 See Niels Bohr, "Discussion with Einstein on Epistemological Problems in Atomic Physics," in *Atomic Physics and Human Knowledge* (1949; New York: Science Editions, 1961), 32–66.

47 Despite his work in Nazi Germany, Heisenberg's relationships with the physicists mostly recovered postwar. Bohr and Heisenberg continued to work and vacation together despite never reaching a "shared view" of Heisenberg's wartime activities. Cathryn Carson, *Heisenberg and the Atomic Age: Science and the Public Sphere* (Cambridge: Cambridge University Press, 2010), 22. Born vilifies Heisenberg as "Nazified" in letters from 1944 to 1948, but he retracted these comments as "probably not justified" in the 1960s, believing that the evidence "justifies [Heisenberg's] behavior during the time." Max Born, *The Born-Einstein Letters: Friendship, Politics and Physics in Uncertain Times*, trans. Irene Born (New York: Macmillan, 1971), 141, 163, 165, 167.

48 For example, Bohr extends the discoveries of quantum mechanics to biology and psychology by analogy. Niels Bohr, "Light and Life," in *Atomic Physics* (1933; New York: Science Editions, 1961), 3–12.

49 Heisenberg, *physikalischen Prinzipien*, 43–4.

50 Brecht, *Werke,* 26:168.

51 Ibid., 21:122.

52 David Engels, "Oswald Spengler and the Decline of the West," in *Key Thinkers of the Radical Right: Behind the New Threat to Liberal Democracy*, ed. Mark Sedgwick (Oxford: Oxford University Press, 2019), 3–21.

53 Carson offers an overview of his war years (and the history of their interpretation) according to recent evidence. She summarizes: "At the extremes, Heisenberg has been portrayed as a man of the resistance who withheld the bomb from Hitler or a German nationalist who would have delivered it if not tripped up by his own arrogance. In between, careful interpreters have still disagreed"; Carson, *Heisenberg*, 22.

54 Ibid., 23.

55 Ibid., 22.

56 Philip Ball, *Serving the Third Reich: The Struggle for the Soul of Physics Under Hitler* (Chicago: University of Chicago Press, 2014), 99–101.

57 Brecht, *Werke*, 26:682.

58 Werner Heisenberg, "Über quantentheoretische Umdeutung kinematischer und mechanischer Beziehungen," *Zeitschrift für Physik* 33 (1925): 879–93. Objects in quantum states have a probability of leaping into a different state. For a popular primer, see Philip Ball, "Quantum Leaps, Long Assumed to Be Instantaneous, Take Time," *Quanta*, 5 June 2019, https://www.quantamagazine.org/quantum-leaps-long-assumed-to-be-instantaneous-take-time-20190605/.

59 See Lukas Mairhofer, *Atom und Individuum: Bertolt Brechts Interferenz mit der Quantenphysik* (Berlin: De Gruyter, 2022).

60 Heisenberg, *physikalischen Prinzipien*, 43–4.

61 Ibid., 18: 229.

62 Heisenberg, *physikalischen Prinzipien*, 49. This is no longer how physicists understand uncertainty. Now, objects are thought to lack fixity because of their wavelike properties, not the interference of measurement.

63 Geiger, Franz, and Scheel, *Handbuch*, 1–2.

64 Bertolt Brecht, "Dänische Fassung 1938/9," in *Leben des Galilei: drei Fassungen, Modelle, Anmerkungen* (1943; Berlin: Suhrkamp, 1998), 27.

65 Ibid., 20.

66 Bohr, "Discussion with Einstein," 39. Emphasis in original.

67 Bertolt Brecht, "Amerikanische Fassung 1947," in *Leben des Galilei: drei Fassungen, Modelle, Anmerkungen* (1943; Berlin: Suhrkamp, 1998), 143.

68 Brecht, "Dänische Fassung," 36.

69 Brecht, *Werke*, 59.

70 Quoted in Bohr, "Discussion with Einstein," 47. Emphasis in original.

71 Ehrhard Bahr, *Weimar on the Pacific: German Exile Culture in Los Angeles and the Crisis of Modernism* (Berkeley: University of California Press, 2007), 103.

72 Brecht, *Werke*, 18:97.

73 Ibid., 26:451.

74 Brecht, "Dänische Fassung," 85.

75 Brecht, *Werke*, 85.

76 Parker, "Taoist Paradox and Socialist-Realist Didactics: Re-Grounding the Galileo Complex in the 'Danish' *Leben des Galilei*, Brecht's Testimony from the 'Finsteren Zeiten,'" *German Life and Letters* 69, no. 2 (2016): 204–5.

77 Brecht, *Werke,* 18:97.
78 Ibid., 3:1125.
79 Ibid., 3:1126.
80 Ball, *Serving the Third Reich,* 99.
81 Stephan Schwarz, "Defending Alignment: Mimetics, Rationalization, and Rhetorical Fallacies Among Physicists in the Third Reich," *Sage Open* 6, no. 2 (2016). doi: 10.1177/2158244016643565.
82 Ibid.
83 Ibid.
84 Brecht, "Dänische Fassung," 119.
85 Ibid., 104.
86 Ibid.
87 Brecht, "Amerikanische Fassung," 199.
88 Ibid.
89 Bertolt Brecht, "Berlin Fassung 1955/56," in *Leben des Galilei: drei Fassungen, Modelle, Anmerkungen* (1956; Berlin: Suhrkamp, 1998), 376–8.

9

"Let Science and Art Have at It"[1]

The Living Newspapers Perform Science to Promote Depression-Era Theatre/ Squonk Performs Theatre to Promote Trump-Era Science

Emily B. Klein

During the 2016 US presidential election, the Federal Theatre Project's (FTP) 1936 stage adaptation of Sinclair Lewis's *It Can't Happen Here* was a touchstone for many theatre historians who found eerie echoes of the 1930s reverberating with increasing contemporary relevance.[2] The script's cautionary tale about a populist president installing a violent fascist regime on US soil was the product of a rush job by Lewis, who whittled down his 1935 novel into a script that could be easily staged by local FTP outposts around the country in the days leading up to a looming national vote. Though Lewis himself said "It isn't that good," when asked whether he thought the play could incite public unrest,[3] he believed in the urgency of the show's message during a time of sustained economic crisis, climate disasters, and deepening xenophobia and political divisions. Noting the uncanny correlations to our own time, the Berkeley Repertory Theatre in California went so far as to write and stage an updated version of Lewis's play in the fall of 2016 as "a reminder that democracies are fragile things."[4]

Since Trump's inauguration, US theatres have continued to use their productions as a way to sound the alarm about the early warning signs of fascism playing out before our eyes, and, like the FTP, many have worked to creatively invigorate audience engagement and extend their reach to new communities during a time when arts budgets have been increasingly imperiled.[5] The FTP's productions emerged during a moment of intersecting national crises that remarkably foreground and parallel the conditions faced by US theatres today; the 1929 stock market crash and the ensuing mass unemployment of the Great Depression, severe droughts throughout the United States that led to the Dust Bowl's hunger and farming emergencies, race riots in major cities, and surging xenophobia inspired by waves of new immigration and refugees fleeing the rise of Hitler and Mussolini are all eerily mirrored in contemporary American culture. With Covid-19 launching an economic downturn that further inflamed ongoing hunger and housing crises in the US, unprecedented climate disasters and catastrophic wildfires, a national protest movement for racial justice to end police brutality, and an international rise in right-wing extremism coinciding with the election of a US president famous for building border walls, banning Muslim travel, and separating refugee families at American entry points, the similarities between the two historical moments are undeniable. And yet, one major difference between then and now can be found in populist approaches to concepts like "truth" and "facts," "science" and "the media."

During the four years that the FTP was in operation under the auspices of the Works Progress Administration (WPA), those terms were not the provocations and slurs of derision that they are in much of the United States today. To the contrary, the FTP's Living Newspaper administrators actually discovered that associating their theatrical endeavors with new forms of science, technology, and journalism helped them frame their performances for reluctant bureaucrats as a worthy new communication medium—one that was factual, masculine, reliable, and scientifically advanced. Though Roosevelt's opponents were eager to slash funding from the WPA budget for anything that struck them as un-American, communist, or frivolous, FTP administrators championed the Living Newspapers as didactic tools for maintaining an informed citizenry. The "small" and "precious" domestic themes associated with parlor dramas were openly eschewed by Living Newspaper leaders who sought to present hard-hitting news in their unique form of fast-paced informational plays for mass audiences.[6] Their short, serialized scenes provided facts on current events from different perspectives. In keeping with contemporaneous new developments in modern psychology, journalism, epidemiology, and industrial chemistry, the Living Newspapers reflected the interests and exigencies of the time by experimenting with early forms of what we now call "documentary theatre," combining elements of the cinematic newsreel and the theatre of information in their mimetic search to represent empirically verified facts onstage.[7] It appears that by rhetorically indexing those fact-based pursuits of science and truth, Living Newspaper administrators briefly won the support of conservative lawmakers eager to cut arts funding from Roosevelt's New Deal agencies.

By contrast, linguistic markers of scientific fact make meaning in dramatically different ways in our current post-Trump and mid-Covid-19 context. During Trump's presidency nearly every mainstream news outlet reported on his proclivity for spreading misinformation and persuading followers that documented facts were merely "fake news" if they were "unflattering" or "don't fit with his world view."[8] Reporting on Trump's 30,573 false claims during his four years in office, a special fact-checking team for the *Washington Post* writes, "What is especially striking is how the tsunami of untruths kept rising the longer he served as president and became increasingly unmoored from the truth."[9] It is tough to overstate the harm that the Trump administration did, in particular, to scientific agencies and popular beliefs about science in the United States. Science reporter Jeff Tolleson writes:

> [Trump's] administration has undermined, suppressed and censored government scientists working to study the virus and

reduce its harm. [. . .] Trump's actions in the face of COVID-19 are just one example of the damage he has inflicted on science and its institutions over the past four years, [he] also backpedalled on efforts to curb greenhouse-gas emissions, weakened rules limiting pollution and diminished the role of science at the US Environmental Protection Agency (EPA). Across many agencies, his administration has undermined scientific integrity by suppressing or distorting evidence to support political decisions, say policy experts.[10]

Thus, in our current era of fake news, anti-vaxxers, a surging anti-science movement, and the slow death of print journalism, this chapter examines how theatrical practices and public discourses about science have informed one another during two moments of social and economic upheaval in the United States—the Great Depression and the Covid-19 pandemic. Undoubtedly, the history of US theatre is also a history of labor precarity and the struggle for resources,[11] but the Great Depression is a particular moment in our national past with an artistic legacy that can be observed in the productions and practices of theatres striving to make new work in the shadow of Trump and the pandemic today. Ultimately, this chapter explores how rhetorics of science and truth briefly bolstered the Living Newspapers and how theatres today are adapting those efforts as they endeavor to salvage popular attitudes about science while keeping writers and actors employed. By bridging our current moment with the Great Depression, this work investigates the value of art in times of crisis and how science has operated over time as a powerfully volatile concept in our national public sphere.

From 1935 to 1939, FTP administrators persuasively advocated for the Living Newspapers as informational tools for promoting not only engaged citizenship but also populist narratives of American scientific progress. This approach helped the FTP gain traction with policy makers who ideologically disavowed funding the performing arts. As one Congressman famously proclaimed, "Culture! What the Hell—Let 'em have a pick and shovel."[12] In a country without a history of socialized arts programs, FTP leaders successfully—if briefly—convinced reluctant policymakers to fund a national public theatre for the first and only time in American history.[13] At their final tally, the FTP's 63,728 performances welcomed 30,398,726 attendees for a mere $46,207,779, "approximately the cost of one complete battleship."[14] As FTP Director Hallie Flanagan wrote, "For this amount an average of 10,000 people supported an average of four dependents for four years."[15] No matter one's feelings about the theatre, the FTP knew it was impossible to deny the labor and employment successes that the numbers revealed.

In comparison, the contemporary resurgence of science-oriented theatre in the United States suggests that as artists and arts administrators across the country have fought to save American theatre—first, from Trump's attempts at massive federal defunding,[16] and now, from collapse during the pandemic—they are creatively inverting the discursive strategies of their Depression-era counterparts by leveraging the performing arts as tools to make public health and science more accessible and less polarizing. Through their innovative Science, Technology, Engineering, Arts, and Math (STEAM) workshops and public performances, a variety of theatre groups explore how scientific discourses are operating in the public imaginary today. Recent shows include *Code Blue* and *Floor Wipers* by Philadelphia's Wilma Theater, the *Covid Monologues* by Baltimore's Single Carrot Theatre, and *Feel the Spirit* by Berkeley's Shotgun Players, while large-scale initiatives like those of Cal Tech and Pasadena Playhouse's collaborative MACH 33 and the Sloan Foundation fund science-oriented theatre by students and producing partners.[17]

One contemporary group of particular interest to this chapter has been operating for the last thirty years in Western Pennsylvania—the same state where the FTP once staged its only East Coast production of the public health play, *Spirochete* (1938), "A Living Newspaper on Man's Conquest of Syphilis." Squonk (formerly Squonk Opera) takes a unique approach to science-oriented performance. They call their work "site-specific and participatory, [from] places where people like their art boisterous and their food greasy."[18] Before the pandemic, they debuted *Hand to Hand* (2019), an experimental outdoor touring show that gives "a big hand for science."[19] Often commissioned for outdoor festivals and civic events, the current show, like most of their recent work, uses the large-scale spectacle of a multilevel, multistory performance platform on which musicians and stagehands perform and interact with one another and the audience. Making their theatrical mechanics and labor intentionally visible is always part of Squonk's aesthetic, as are the audience interactions that are built into each show. In *Hand to Hand*, the platform structure is flanked by two giant inflatable purple hands to tell a musical tale about the power we all hold (quite literally) in our hands to communicate, help, build, and destroy. Originally titled *Grab* in reference to the leaked recording of Trump boasting about his predatory assault habits, the show was first developed in response to his agenda of misinformation and discrimination but grew to take on a more expansive and hopeful approach, like much of Squonk's repertoire.[20] Like the FTP, Squonk often funds their performances through federal grants, but in contrast with the Living Newspapers, which openly used scientific language to protect and promote theatre, Squonk uses theatrical performance as a device to subtly champion scientific discourses at free public shows, even visiting Title 1 elementary schools to conduct experiments onstage before the pandemic began. Thus, in the eighty years between the US surgeon general's endorsement of the FTP's *Spirochete* and the debut of this new work by Squonk, we witness an inversion of sociocultural exigencies: rather than advancing theatre through the FTP's thematic and linguistic indexing of science, this contemporary group is using theatre to change attitudes and advance public discourses about science.

Legitimizing Theatre in the 1930s

Spirochete: A Play about the Power of Science to Stop an Epidemic

While theatre historians have cogently analyzed the FTP's Depression-era scripts, production histories, and promotional artwork,[21] its trove of well-archived administrative and outreach documents has not received the same scholarly attention. This oversight has allowed us to miss the significant ways that administrative FTP materials inventively echo the civic and informational themes of their plays. In effect, FTP leaders were pioneers in the field of social marketing, attempting to influence public health behavior long before the field was established.[22] The FTP's discursive turn to the rhetorics of science and fact was informed by their understanding of theatre's necessary reinvention as a tool of public information during a time of economic crisis.[23] This theme is evident in an audience report from February, 1936, written by Play Bureau administrator Larry Harr:

"We need an American National Theatre just as we need schools, colleges and libraries. Some people learn in the theatre more than they learn in school. There are many people who read no books, but they go to theatres."[24] In keeping with this didactic approach, regional Living Newspaper offices were structured not like theatre companies but daily city papers, "run by an editor-in-chief, managing editor and city editor, who worked with reporters and copyreaders to dramatize the news."[25] FTP supervisor and founder of the American Newspaper Guild, Heywood Broun, saw this practice as essential for portraying theatre-making as decent work and theatre-going as the duty of an informed citizenry, equivalent to reading the news. Including journalists and fact checkers within the play development process also helped improve public perceptions of the Living Newspapers' legitimacy, as did their regional offices' specialized attention to local data points on public health, like the regional statistics on housing in *One-Third of a Nation*—a drama interested in urban overcrowding—and infection and mortality rates in *Spirochete*—the FTP's famous syphilis prevention play.[26]

Spirochete: A Living Newspaper on Man's Conquest of Syphilis was written by Arnold Sundgaard in 1938 under the supervision of famed bacteriologist and medical science writer Paul de Kruif and US surgeon general Thomas Parran, whose book *Shadow on the Land* (1937) detailed his lifelong work to curb the disease's spread. Also aided by the mentorship of playwright Susan Glaspell, Sundgaard developed a two-act script that moved from Columbus's role in intercontinental disease transmission in the 1490s to the present-day epidemic of the 1930s. Given Parran's personal commitment to this cause and his estimate that some 680,000 people were under treatment for syphilis while 60,000 babies were being born annually with congenital syphilis, *Spirochete* became part of his initiative to tackle the epidemic by increasing support for testing and treatment at the state level.[27] His National Venereal Disease Control Act, which was passed by Congress in 1938, was the first coordinated federal response to syphilis and the only one to be promoted by a national theatrical production.

The play's structure followed the conventions of the Living Newspaper genre with the Loudspeaker (sometimes referred to as the "Voice of the Living Newspaper") swiftly guiding the audience through a series of brief historical vignettes. Beginning in the present day with the introduction of a young couple applying for their marriage license, the scene's characters served as anchors for audience identification since their sheepish curiosity launched the play's multicentury exploration of syphilis' spread from Columbus's crew to new shores. The chronological episodes eventually returned the audience to the present day where they witnessed workers, employers, and politicians debating the best way to handle the epidemic. Ultimately, like most Living Newspapers, the play ended with a call to action that broke the show's fourth wall: "This fight must go on until syphilis has been banished from the face of the earth. It can be done and will be done if you and you and you wish it so. The time has come to stop whispering about it and begin talking about it . . . [sic] and talking out loud!" (Figure 5).[28]

FIGURE 5 *The Federal Theatre Project performs* Spirochete *in Seattle, 1939 (courtesy of the University of Washington Libraries, Special Collections Division, UW Theatres Photograph Collection (PH Collection #2360, box 4, folder 17)).*

While the initial production run was well attended and critically praised in Chicago, the FTP struggled to market a topic so taboo that the word "syphilis" had not even appeared in print media until the previous year—1937—and many audience members had never heard it spoken in public.[29] Although the discovery of an effective antimicrobial remedy for syphilis, Salvarsan, was almost thirty years old when *Spirochete* premiered, the disease was shrouded in so much silence and shame that it was difficult to increase public awareness about prevention and treatment. "Many still assumed that the disease was a punishment for sin (a view represented by the character of the Reformer in Spirochete, who is angrily expulsed from the research laboratory), or that it was only contracted by the lower classes, 'perverts,' or African Americans."[30]

The US government's infamous Tuskegee Experiment, which began in 1932, only helped to perpetuate the racist stigmas and pseudo-scientific beliefs associated with the disease. The Public Health Service's unregulated forty-year study on the effects of untreated late-stage syphilis in 399 Black male subjects was "a nontherapeutic experiment" that "had nothing to do with treatment"; medication, information, and even the syphilis diagnosis were all withheld from patients with the singular goal of observing the severity of complications caused by the disease's unchecked progression.[31] Since being revealed by a whistleblower in 1972, the study has been internationally condemned as perhaps the longest-running bioethics violation in modern medical history. Meanwhile, the Depression-era federal spending in communities deemed worthy of public health information, testing, and treatment was still seen in some circles "not only as an attack on decency, but also as government infringement on private medicine," since doctors could get away with charging up to $25 for confidential syphilis treatment as compared to a typical $3 office visit for most other routine care.[32] While public clinics offered more affordable options, they lacked the same reputation-preserving privacy as expensive private providers. Thus, the history of syphilis in the United States is not only a history of tensions between public and private healthcare funding, but also a history of federal approaches to medical research and public health being alternately at odds with individual safety, private sector profit, and social taboo.

Hallie Flanagan alluded to the controversies that followed *Spirochete* from one regional production to the next: "To handle this theme clearly and directly took courage, for while these are not the days of Brieux or Ibsen, still it is a hazardous undertaking to trace the history of the most deadly of all social diseases, to show its insatiable spread over the earth, and to recount the unremitting battles of scientists to isolate the germ and effect the cure."[33] After a successful opening run in Chicago *Spirochete* opened in Seattle, where medical groups hesitated to sponsor it, fearing it could stoke public fears about mandatory syphilis testing. In Philadelphia, religious leaders tried to close it down and demanded sweeping rewrites omitting "slanderous" references to Columbus and his crew as the original intercontinental carriers of syphilis.[34] Both shows went on anyway, along with additional productions in Portland and Cincinnati, cosponsored and endorsed in each city by local medical organizations, scientists, judges, and newspapers. Often, the productions were also used to rally support for state-level health legislation and low-cost testing. For example, at a mobile testing unit in the foyer of Seattle's Metropolitan Theatre local leaders were invited to take blood tests on-site to destigmatize the practice of getting tested, and in Chicago the play's lobby blood-testing clinic was credited with aiding in the discovery and treatment of 56,000 new cases of syphilis in that city alone.[35]

The use of what we might consider modern social marketing and arts publicity concepts like cross promotion, audience segmentation and outreach, and behavior determinants and barriers was foundational for attracting audiences to FTP shows as well as discursively positioning the FTP for federal decision makers. Evidence of these efforts can be found in the oral history recordings from interviews conducted in the 1970s with FTP publicists, production managers, and other administrators. New York City FTP administrative director Philip Barber and Living Newspaper unit publicity director Ethel Aaron Hauser described their enduring innovations in FTP audience outreach:

> PHILIP BARBER (PB): I still say one of the few things that has survived into the theatre today that you set the pattern for was that audience organization. We did it before anybody else.
>
> ETHEL AARON HAUSER (EAH): Yes, we got unions in, we got unions to come in groups. We got organizations, ladies' groups,

clubs . . . the Ladies Garment Workers Union. We used to encourage them to take a bus and come. We really needed to plug for the audiences.

PB: This had never been done before 'til Ethel did it.

EAH: You know, I never thought of it that way. [. . .] We did promotion pieces to reach them like the one on *Power* and the one on *Injunction Granted*. We sent out legal looking mailings, you know, to get people. We did a lot of direct mail to get them in. And at that time in the theatre it wasn't considered de rigeur, you know.[36]

Regional promotions departments followed the outreach methods established by publicists like Hauser. In addition to the practice of organizing "theatre parties" with discounted tickets for community groups, local Living Newspaper administrators also did outreach to potential co-funders whose interests were advanced by relevant shows.[37] *Spirochete*, for instance, was sponsored in Oregon by the Oregon State Medical Association, the Oregon Social Hygiene Association, the Visiting Nurses' Association, and the city health officer.[38]

Although the FTP was shut down before *Spirochete* could reach all the cities where it was intended for production, scholars have noted its positive public health impact in states where it was performed. One month after its run in Seattle, the state of Washington successfully passed bills 373 and 374 requiring prenatal and premarital testing for syphilis, and after the Philadelphia production, a similar bill passed in Pennsylvania in 1939.[39] Undoubtedly, *Spirochete* helped to change public attitudes toward the disease, which led to a decades-long decline in syphilis cases in the United States. Only recently has a dangerous mixture of decreased public health funding, government mistrust, and Covid-19-related impacts on medical access and resources started to lead to a resurgence of syphilis cases in the United States.[40] As public health reporter Caroline Chen writes, "The consequences of the political nature of public health funding have become more obvious during the coronavirus pandemic"[41]—an understatement that only scratches the surface of the conditions that theatre companies are working to address today.

Legitimizing Science in the 2020s

Hand to Hand: A Play about the Power of Shared Discovery to Promote Science

Though the sciences have never operated independently of the cultures in which they are practiced, the artificial divide between art and science as critical realms of knowledge production and exchange is evident in documentation from the FTP as well as theatre organizations currently operating in the United States. Of course, science and theatre *both* call us to look more closely and to know the world better through formalized methods of observation. *Theatron*, the Greek word from which theatre is derived, literally translates to "looking place," and science is often iconized through the symbols of heightened vision—the microscope, the magnifying glass, and the telescope. As Alan Bleakley and other scholars of the medical humanities have noted, "truth turning" in performance practices including verbatim, testimony, and documentary theatre have led to a "re-*visioning* of medical education" for both scientists, themselves, and the public.[42] Unable to resist the optical metaphor, even in his description of the turn toward information theatre, Bleakley suggests that the mirror of theatre helps researchers and citizens alike, as it is used not only to train public health practitioners for patient interactions but also to teach the public to see scientific information more clearly.

Inviting the public to see science more clearly has become a central goal of Pittsburgh, Pennsylvania-based performance group Squonk. Several of their recent shows, including *Hand to Hand* (2019), its STEAM workshop counterpart *A Big Hand for Science*, and their newest production *Squonk in the Neighborhood* (2021), take up scientific processes of "shared discovery" with the goal of "making a community of the imagination."[43] Just before the pandemic, the nationally acclaimed group performed *Hand to Hand* at the Kennedy Center. Using story-high inflatable purple puppets of disembodied human hands built to bend and move with shocking physiological accuracy, this experimental outdoor show puts the physical and affective elements of human connection under the microscope. With pullies, levers, and cords that do the work of oversized joints, bones, and muscles, the giant hand props become the vehicles through which audiences are prompted to consider both our universal strength and vulnerability as sharers in the human condition. "When we started to make this it was the middle of the Trump years," Squonk cofounder Steve O'Hearn explains, "and we were in a very dark place, emotionally, and struggling with a feeling of impotence that humans generally have and American voters certainly have as they get disenfranchised. So, we thought, how do you describe that in visual terms? And we thought as artists and musicians, we use our hands."[44] As audiences watch the hands speak through gesture alone, the show lets viewers witness a larger-than-life demonstration of how hands, as synecdochal stand-ins for humans themselves, can initiate conflict as easily as connection. The "come hither" curl of an index finger, the approval of a thumbs up, and the affirmation of a high five are juxtaposed in different musical scenes with the negation of a flat-palmed stop and an aggressive thumb war. Over the course of the show and between some of the gestural vignettes, the cast invites audience members to join them onstage to try operating the hand joint rigging themselves. The performers explain that just like a human hand with tendons and joints, the pneumatic finger puppets require both strength and collaboration to move (Figure 6).

A natural extension of the show, the corresponding STEAM workshop makes these lessons even more scientific and concrete:

> *A Big Hand for Science* [is] about dynamic structures in nature and machines, everything from bee wings and finger muscles, screw jacks and levers. How things move and do what they do. The study of our hands, which shape our world with their network of tendons and bones, and opposable thumbs, branches out to geometry and biology, art and engineering, and proving evolution without even saying the word "evolution."[45]

Like the FTP, Squonk is as intentional about using art to teach science as it is about tiptoeing around the political buzzwords that could incite backlash from funders or audiences. For the FTP, their focus on science helped to (temporarily) ward off accusations of communism and un-Americanness; for Squonk, their focus on science is the very thing that causes the group to risk their Americanness being called into question.

FIGURE 6 *Squonk performs* Hand to Hand *at PPG Plaza in Pittsburgh, Pennsylvania, in 2019 (courtesy of Squonk, photo by Emily O'Donnell).*

Since their start in 1992, Squonk has been the recipient of nine grants from the National Endowment for the Arts, in addition to receiving support from dozens of other state, city, and private agencies including the Pennsylvania Council on the Arts, the Pittsburgh Foundation, the Heinz Endowments, and the Jim Henson Foundation. But performing shows about science and having access to federal funding are just two of the many features that connect Squonk and the FTP across their historical divide. Crucially, both groups have written about their commitment to the local and their belief that the sponsorship, endorsements, and co-participation of local groups are essential to the success of the pro-science social marketing work that inheres in their performances. As O'Hearn attests in a 2021 grant narrative about *Squonk in the Neighborhood*, which combines their science focus with new mid-pandemic exigencies, "Squonk engage[s] Pittsburgh general audiences with regional grassroots events, forging more continuous and intimate connections with our home communities, without excluding them because of cost."[46] Through collaborations with local artists, high school marching bands, community drumline performers, and Boys and Girls Clubs, each outdoor Covid-19-conscious *Squonk in the Neighborhood* production to date has been tailored to feature the residents and artists of each neighborhood locale.

As if channeling the nascent social marketing instincts of Hallie Flanagan, O'Hearn echoes her impulse to incorporate local community members into the work of each Living Newspaper to make public health messages seem more homegrown, familiar, and relevant. In 1938, Flanagan touted the importance of "an exhibit of paintings by children of the slums" in the lobby of New York's Adelphi Theatre where *One-Third of a Nation* was being performed to raise awareness about the public health crisis caused by overcrowded housing.[47] "The pictures are a part of the play, the play a continuation of the pictures, and both at once a part of the life of the audiences pouring nightly into the Adelphi and a force galvanizing that audience to some sort of action."[48] Similarly, Squonk explains that their new work galvanizes communities through a combination of performing the local and the factual:

> *Squonk in the Neighborhood* will address the pandemic's lockdown isolation, the rise of anti-science conspiracies, and a clear demand to address equity issues. It is vital, now, that we contribute, through *Squonk in the Neighborhood*, to a rebirth of public life, critical thought, and joy, in all the corners of Pittsburgh [. . .] after a long pandemic and conspiracy-fueled battle against truth, science, and empathy. [. . .] We hope to unite diverse citizens into communities of the imagination. We hope to celebrate science and art and create a vibrant public life with a post-pandemic celebration that impacts the civic space and inspires connection.[49]

Although these recent internal documents are more explicit about the group's focus on promoting "truth, science, and empathy" than they were under Trump's presidency, they also reflect an increased urgency and an awareness of the damage that has been done to civic life, not only by the coronavirus but also by four years of living under a presidential administration that advanced "anti-science conspiracies."[50] "Our hope is to encourage the re-blossoming of our public space," they venture; "post-pandemic, the strangeness of pedestrian and civic life has increased even more. We want audiences to participate in an optimistic and visceral conversation that we believe is more powerful than preaching. Division, and t-shirt and hat logos, have riven what used to be community spaces into scary battlegrounds,"[51] but spectacle, humor, and information, they argue, can help to return us to a more humanistic form of public life.

Both Squonk and the FTP position theatre as an essential public service, an instructive and informational resource that fills in the gaps of our flailing public education system. Just as Larry Harr defended the FTP by arguing that "Some people learn in the theatre more than they learn in school,"[52] Squonk insists: "Our programs directly attack the archetypal arts and science inequity of public schools, a source of the widening gap between advantaged and disadvantaged. Both of Squonk's Artistic Directors, Jackie and Steve, went to regional public schools. We often engage youth who have little arts or science exposure to develop critical skills and creativity to become engaged citizenry."[53] Their *Big Hand for Science* workshop that toured local schools in 2019–20 was abruptly halted because of pandemic school closures, but it garnered plenty of positive audience feedback before the disruption. In one thank you note to the group, a child named Jake writes, "I didn't know science could be so funny!" and another child, Camryn, proclaims, "Because of this assembly, one day I would like to be a scientist."[54]

Putting *Hand to Hand* and *Spirochete* in conversation pushes us to reconcile the cultural shift that has taken place in the intervening years between the syphilis epidemic and the Covid-19 pandemic, the Great Depression and the economic downturn of 2020. While the endorsements of the surgeon general and local doctors were used as PR for *Spirochete*, it's hard to imagine how the endorsements of scientists, doctors, or epidemiologists would be received by Squonk's audiences around Western Pennsylvania, or in Tulsa, Fort Worth, or Des Moines where they've recently toured the show. While conservative lawmakers of the 1930s were willing to tolerate funding theatre in part because of the Living Newspapers' scientific, factual, and utilitarian purpose, those commitments may be viewed as liberal "dog whistles" today. The organizations that contract and collaborate with Squonk clearly believe in the value of public art and community engagement, but the path from holding those values to openly supporting democratic participation, the social contract, science, and public health is treacherous in the midst of a deeply politicized national health emergency defined by Trump-era anti-science attitudes.

The history of science-based theatre tells us more than we might realize about our local communities and national culture. Not only does it reveal core values of theatre makers, funders, and audiences, but it also tells us something about the heavy anxieties that lie beneath each moment's swirling zeitgeist. In the 1930s polite society, decency, and propriety showed up as cultural concerns in many of the Living Newspapers' plays and reviews. Theatre was widely regarded as too polite and effete, and yet when reimagined as factual and rugged, theatre was effectively used to combat the respectability politics that had stifled syphilis education and prevention for centuries.

By contrast, human connection, shared spaces, and community outreach seem to be popular contemporary concerns that appear in Squonk's marketing language, as well as the social marketing and development work of so many science-oriented theatres. These discourses have roots in the anxieties that O'Hearn mentions earlier: nearly impenetrable political bubbles and political divides are the twenty-first-century outgrowths of old American systems of violence and exclusion. Exacerbated by social media, the Trump administration, and the pandemic, the problems of civic engagement, public discourse, the common good—even getting humans together—have taken on a new urgency in the theatre.

Another concern of this project is the tendency of contemporary critics to treat the FTP's advent as a uniquely isolated event, a brief artistic manifestation of Roosevelt's Works Progress endeavors, the likes of which shall not be seen again. Yet, US theatre-making in the years leading up to and during the pandemic has borrowed formal artistic strategies as well as promotional and rhetorical techniques from the FTP, with today's artists renewing their calls for "a New Deal for the Arts."[55] After the United States spent 2020 lurching into an economic downturn that rivaled the severity of the Great Depression, this study also elucidates some of the questions that bridge our own historical moment with the FTP's—two instances when the arts, culture, and entertainment sectors have faced closures and unemployment severe enough to permanently alter the future of the industry. While teaching the public about science in the 1930s ultimately was not enough to save the Federal Theatre, the spectacle, communitas, and *joy* of live theatre today could have the power to reassure pandemic-fatigued Americans about the value of curiosity, facts, and science to safeguard our future.

Acknowledgments

My thanks go to the organizers and participants of the American Theatre and Drama Society's panel on "Working Conditions, Labor, and Equity in Theatre and Performance Studies" at the Modern Language Association's 2023 Convention for their feedback on portions of this project. This research was supported by the Provost's Faculty Research Grant of Saint Mary's College of California.

Notes

1 "About," Squonk, https://squonk.org/about.
2 See, for example, Ann Elizabeth Armstrong and Joan Lipkin, "'The Every 28 Hours Plays' and 'After Orlando': Networked, Rapid-Response, Collective Theatre Action—New Forms for a New Age," *Theatre Topics* 28, no. 2 (2018): 159–64.
3 Laura Collins-Hughes, "'It Can't Happen Here' Review: A Not-So-Subtle Slide into Autocracy," *New York Times,* 26 October 2020, https://www.nytimes.com/2020/10/26/theater/it-cant-happen-here-review.html.
4 Jay Barmann, "'It Can't Happen Here' At Berkeley Rep Is a Timely if Heavy-Handed Political Cautionary Tale," *SFist,* 1 October 2020, https://sfist.com/2016/10/01/it_cant_happen_here_at_berkeley_rep/.
5 See Michael Cooper et al., "Arts Groups Draft Battle Plans as Trump Funding Cuts Loom," *New York Times,* 19 February 2017, https://www.nytimes.com/2017/02/19/arts/nea-cuts-trump-arts-reaction.html.
6 See my previous research on masculinity (202–4), journalism (204–10), and labor precarity in the Federal Theatre, which foregrounds a few of the ideas in this chapter: Emily Klein, "'Danger: Men Not Working,' Constructing Citizenship with Contingent Labor in the Federal Theatre's Living Newspapers," *Women & Performance: A Journal of Feminist Theory* 23, no. 2 (July 2013): 193–211.

7 Scholars have traced the roots of documentary theatre to ancient Greece as well as early cinema. See, for example, Attilio Favorini, *Voicings: Ten Plays from the Documentary Theatre* (New York: Ecco Press, 1995); and Jane M. Gaines, "Radical Attractions: The Uprising of '34," *Wide Angle* 21, no. 2 (March 1999): 101–19.

8 Jane C. Timm, "Trump vs. the Truth: The Most Outrageous Falsehoods of His Presidency," *NBC News*, 31 December 2020, https://www.nbcnews.com/politics/donald-trump/trump-versus-truth-most-outrageous-falsehoods-his-presidency-n1252580.

9 Glenn Kessler et al., "Trump's False or Misleading Claims Total 30,573 over 4 Years," *Washington Post*, 24 January 2021, https://www.washingtonpost.com/politics/2021/01/24/trumps-false-or-misleading-claims-total-30573-over-four-years/.

10 Jeff Tolleson, "How Trump Damaged Science—and Why It Could Take Decades to Recover," *Nature*, 5 October 2020, https://www.nature.com/articles/d41586-020-02800-9.

11 Klein, "'Danger,'" 193–8.

12 Joanne Bentley, *Hallie Flanagan: A Life in the American Theatre* (New York: Alfred A. Knopf, 1988), 340.

13 Charlotte M. Canning, *On the Performance Front: US Theatre and Internationalism* (London: Palgrave Macmillan, 2015), 26–8, 100–9.

14 Hallie Flanagan, *Arena: The Story of the Federal Theatre* (New York: Limelight Editions, 1985), 435–6.

15 Ibid.

16 For details on four years of Trump's annual federal budget plan attempts to eliminate the National Endowment for the Arts, the National Endowment for the Humanities, and the Corporation for Public Broadcasting, see Peggy McGlone, "Trump Budget Again Calls for the Elimination of Federal Arts Agencies," *Washington Post*, 10 February 2020, https://www.washingtonpost.com/entertainment/trump-budget-again-calls-for-the-elimination-of-federal-arts-agencies/2020/02/10/8b9e8df2-4c4f-11ea-bf44-f5043eb3918a_story.html.

17 For other examples of contemporary theatre about public health, the pandemic, climate change, and environmental racism, see the work of Superhero Clubhouse, http://www.superheroclubhouse.org/; the awardees of the Alfred P. Sloan Foundation, https://sloan.org/programs/public-understanding/theater; and B. Appiah, B. Walia, and S. H. Nam, "Promoting COVID-19 Vaccination through Music and Drama—Lessons from Early Phase of the Pandemic," *British Journal of Clinical Pharmacology*, 18 August 2021, 1–4.

18 Squonk, "About."
19 Squonk, *A Big Hand for Science* S.T.E.A.M. Workshop, Audience Feedback Packet, 2019, https://squonk.org/hand-to-hand.
20 Interview with Squonk cofounder, Jackie Dempsey, Pittsburgh, PA, 3 July 2022. See also "Transcript: Donald Trump's Taped Comments About Women," *New York Times*, 8 October 2016, https://www.nytimes.com/2016/10/08/us/donald-trump-tape-transcript.html.
21 For example, see Rena Fraden, *Blueprints for a Black Federal Theatre 1935-1939* (Cambridge: Cambridge University Press, 1994); Elizabeth A. Osborne, *Staging the People: Community and Identity in the Federal Theatre Project* (London: Palgrave Macmillan, 2011); Susan Quinn, *Furious Improvisation: How the WPA and a Cast of Thousands Made High Art out of Desperate Times* (New York: Walker & Company, 2008); and Barry B. Witham, *The Federal Theatre Project: A Case Study* (Cambridge: Cambridge University Press, 2003).
22 Philip Kotler and Gerald Zaltman, "Social Marketing: An Approach to Planned Social Change," *Journal of Marketing* 35, no. 3 (July 1971): 3–12.
23 Witham, *The Federal Theatre Project*, 78.
24 Larry Harr (Leo Schmeltsman), "The Audience of the Federal Theatre." Library of Congress: Federal Theatre Project Archives, Box 960: folder 2, 3.
25 Bonnie Nelson Schwartz, *Voices from the Federal Theatre* (Madison: University of Wisconsin Press, 2003), 58.
26 See Morris Watson, "The Living Newspaper: The Federal Theatre Dramatizes the Events of the Day," *Scholastic*, 31 October 1936, and "Playbill from production of *Power*" ("Living Newspaper," 1, 3). Library of Congress, Music Division, Federal Theatre Project Collection, Box 1096.
27 Caroline Chen, "Syphilis Is Resurging in the U.S., a Sign of Public Health's Funding Crisis," *NPR, Morning Edition*, 1 November 2021, https://www.npr.org/sections/health-shots/2021/11/01/1050568646/syphilis-std-public-health-funding.
28 Arnold Sundgaard, *Spirochete*, Federal Theatre Project, National Service Bureau (Publication No, 74-S, January 1939), 114.
29 See Sarah Guthu, "Living Newspapers: Spirochete," *The Great Depression in Washington State Project,* 2009, https://depts.washington.edu/depress/theater_arts_living_newspaper_spirochete.shtml.

30 Ibid.
31 James H. Jones, *Bad Blood: The Tuskegee Syphilis Experiment* (New York: The Free Press, 1981), 2.
32 Guthu, "Living Newspapers: Spirochete."
33 Flanagan alludes to the stigma of syphilis represented in Henrik Ibsen's *Ghosts* (1882) and *A Doll's House* (1879) and Eugène Brieux's *Les Avariés* (1901) or *Damaged Lives*. See Flanagan, *Arena,* 144.
34 Arthur R. Jarvis, Jr., "The Living Newspaper in Philadelphia, 1938-1939," *Pennsylvania History: A Journal of Mid-Atlantic Studies* 61, no. 3 (July 1994): 345.
35 Quinn, *Furious Improvisation*, 233.
36 Ethel Aaron Hauser and Philip Barber, Interview transcript with Lorraine Brown and Diane Bowers, February 20, 1976, Series 1: Transcripts 1961-1984, Box 5, folder 19, Works Progress Administration oral histories collection, C0153, Special Collections and Archives, George Mason University.
37 Jarvis, "The Living Newspaper in Philadelphia," 348.
38 Flanagan, *Arena,* 301.
39 See Guthu, "Living Newspapers: Spirochete" and John O'Connor and Lorraine Brown, *The Federal Theatre Project: Free, Adult, Uncensored* (London: Methuen, 1986), 98.
40 Chen, "Syphilis Is Resurging."
41 Ibid.
42 Alan Bleakley, "Foreword," in Alex Mermikides, *Performance, Medicine and the Human* (London: Bloomsbury Publishing, 2020), 8. Emphasis mine.
43 "*Hand to Hand*," Squonk, https://squonk.org/hand-to-hand.
44 Scott Mervis, "Squonk Opera gets its groove back with 'Hand to Hand,'" *Pittsburgh Post-Gazette*, 13 July 2021.
45 Steve O'Hearn, *Squonk in the Neighborhood* (Grant Narrative, 2021), 7.
46 Ibid., 4.
47 Hallie Flanagan, "Theater and Geography," *Magazine of Art*, August, 1938, Library of Congress: Federal Theatre Project Archives, Box 960: folder 2, 1.
48 Ibid., 1.
49 O'Hearn, *Squonk in the Neighborhood*, 1, 6.

50 Ibid., 1.
51 Ibid., 7.
52 Harr, "The Audience," 3.
53 O'Hearn, *Squonk in the Neighborhood*, 7.
54 Squonk, *A Big Hand for Science* S.T.E.A.M. Workshop, 2.
55 Soraya Nadia McDonald, "A COVID-19 Vaccine Is Here, but Theaters Seek a New Deal," *Andscape*, 30 December 2020, https://andscape.com/features/a-covid-19-vaccine-is-here-but-theaters-seek-a-new-deal/. See also Jerald Raymond Pierce, "So What Could a New Federal Theatre Project Actually Look Like?" *American Theatre*, 3 February 2021, https://www.americantheatre.org/2021/02/03/so-what-could-a-new-federal-theatre-project-actually-look-like/ and Jeremy O Harris, "American Theater May Not Survive the Coronavirus. We Need Help Now," *The Guardian*, 25 January 2021, https://www.theguardian.com/commentisfree/2021/jan/25/american-theater-coronavirus-federal-help.

Creative Interlude

From *Variation for Three Voices on a Letter to Nature*

Diane Stubbings

Note on the Text

Variation for Three Voices on a Letter to Nature is a performance text for three (female) voices that emerged from practice-based research into the application of biological processes to the generation of writing for performance. My aim was to model DNA replication and evolutionary development such that developmental biology was embedded not just in the text's imagery and themes but also in its fundamental dramaturgy.

Beginning with two initiating texts—the Watson and Crick letter to *Nature* (1953), which first set out the structure of DNA, and a passage from Samuel Beckett's *Krapp's Last Tape* (1958)—the foundational process of DNA replication which engendered *Variation* was enacted through textual transcription (the typing and retyping of the text) over six iterations. Responding to visual representations of DNA replication, which show one DNA strand breaking apart before exploiting the chemistry of the cellular environment to reform as two separate strands, I allowed the text to expand and mutate through an accumulation of typing errors and the intrusion/insertion of memories and randomly sourced fragments from other texts.

Through this process, for example, the poet Clive James's reflection that every human life is its own poem was assimilated into one of the iterations. In subsequent iterations, this reflection caught at images of Virginia Woolf's suicide, of her fear (noted in her diaries) that she would drown in other people's words, that there were creatures in her head that needed to be written. These images, in turn, caught at the description of a vulvaless worm I encountered in my research, the worm eventually eaten by its own progeny.

Biological development is dependent on a complex interplay of structural, organizational, and environmental forces. In dramaturgical terms, a performance text generated according to a biological model of development is contingent upon

- the physical, cultural, and imaginative environment within which the text is nurtured;
- the structural forces that arise through the reflexive associations to which the words, phrases, and images with the text give rise as it is being copied; and
- the organizational forces that flow downward from the "author" of the work.

Essential to this dynamic is the equilibrium that is maintained between these forces. This equilibrium may be subverted depending on the biological process being modeled—for example, a cancerous dramaturgy demands that structural forces dominate—but what is crucial to a biological dramaturgy is the eschewing of any authorially imposed form, design, or meaning. Rather, form and meaning emerge and evolve from within the developing work itself.

*

VOICE 2:

> The river flows out of her like a poem, like a life spoken but not yet heard
>
> Dead and gone now, but the stories pressing through like illness, burying themselves in the deep dark blood of her—in the deep dark red of her—the shadows of the room out of which he allowed himself to be born

Breathe

Embryologists have always wanted to know the cell lineage of an organ—for example, do all cells that make the heart descend from one or more founding cells? Do all the thoughts that make the soul descend from one or more founding thoughts

Ensoulment at conception

And the pain? The enduring pain? When does that begin?

Breathe

Some eggs have what it takes to support the embryo's development past the eight-cell stage and some don't. That's all there is to it

A character's way of making another character

The hour hand on the clock, the water quiet beneath us, the river staunched, lazy circles atop the

I held it inside of me, but only for a little time

The question being why, for her, immersion is death, and why, for her, immersion is the very source of life, the particles and pieces of death finding their momentary pattern

She feels herself there, a body under the water, being dragged from the bottom of the river

And she wonders

Angle of incidence equals angle of reflection

Breathes

What was she thinking, as she walked to the river that day?

So little of her left now

There are two chairs in the room. They sit one either side of the bed. There is a rift

How cold the water felt against my legs, it rising up and over the lips of my shoes, and the instinct to take a step back as I felt it beginning to tighten around my ankles, the water, beginning to nestle against the soles of my feet like a pair of slippers I'd forgotten to warm before the fire

The soul, the imagination, the thing-ness of it, the embryo—it is all the mother, all the mother, until it is born

I dozed and drowsed and seemed to feel the sun in my brain seeking all my thoughts

And what is born? And who's to say whether it is alive or dead, and whether breath alone is enough for life and cold alone sufficient for death

The broken branches, the putrid cluster of histories torn from the banks of the past, the muck of so much said, the echoes of lives long passed

Breathe

Breathe

Breathe

He died today. A famous man

They are all dead now

Anticipating the meagre rise and fall of the yellow sheets

They live in the loopholes of natural laws, seeking extensions, exceptions, excuses

Sexual reproduction demands the collapse of the organism

This is the noise she creates in her head

A cell avoids death by dividing. Sacrifices its own unique existence by becoming two

Time standing still and them turning lazy circles

She lay stretched out on the floorboards

In his dreams she did

Her hands under her head

Her hands drifting just under the surface

Him with his tongue hanging on the rim of his lips

The gyandomorphs show that the first plane of cleavage is random

She cannot find them, the words in her head, the wound on the inner flesh of her thigh

The eggs laid by the vulvaless worm have no way out of their mother's womb, and the worm is, as a result, swallowed up, swallowed alive by its own unborn

Words

Telling her what to become

Living is the destruction of every voice, of every point of origin. Living is that neutral composition, oblique space, where our self slips away, the negative where all identity is lost, starting with the very identity of the body living. It is only then, when the body enters into its own dying, that it begins to live

Who to be?

The sea whispering along the river, whispering towards me, leaning in, and there being no sense within me but to lean back into it, walk in, wade in, greet it

Every gene in the chronicle of her cells

Picking them off like stitches

Sitting one either side of death

When his fingers unlace you

The bases are on the inside of the helix, on the outside, the indifferent universe

If I could shed myself of myself, leave nothing but the definition, the contours of something worthwhile

The structure is an open one

What she was trying to drown, really, was the noise

The blood, the clag, the loss, the relief, the life wrapped inside it, and the life wrapped inside that, and the life wrapped inside that, and all the way towards eternity

Does anyone see her hand shaking?

Breathe

CREATIVE INTERLUDE

It was as though I felt myself there already—already dissolved—felt myself flowing there around my feet, pressing the old broken parings of my death into the gaps left of life

Her hands all wet and slimy from the water

Hand to hand to lip to hand

Yours as ever

Life never does more than imitate living, and the living itself is no more than a tissue of trials and errors, a fabrication of all that has been lost to the past and the future

It has been found experimentally

The cells that form the first wild honey, her standing as a child in a fur-lined coat, a knitted embryo clasped in her hand, a solid rod of cells transformed into a hollow cambric thirst-fevered tube

Flowing around my feet, pressing into my broken cells

Every human life its own epic poem

It has been found experimentally

My heart sinking like a shattered boat

It has been found experimentally

The white rush of his words

It has been found experimentally

The silent wall of them breaking

It has been found experimentally

Not like a wave

It has been found experimentally

Like a crack in the shell of an egg.

The Catastrophist Artists' Roundtable

William DeMeritt, Martine Kei Green-Rogers, and Lauren Gunderson[1]

In this artists' roundtable, conducted on Zoom on July 15, 2022, playwright Lauren Gunderson, dramaturg Martine Kei Green-Rogers, and actor William DeMeritt reunited to speak about their collaborations on The Catastrophist, *a one-person play written and filmed as a virtual streaming production during the Covid-19 pandemic lockdowns. Metatheatrical and intimate,* The Catastrophist *places famed virologist (and Gunderson's husband) Nathan Wolfe center stage, where he articulates to his invisible audience—those Nathan perceives in the empty re and those watching on screens from home—his thoughts about science, politics, religion, the birth of his sons, his father's death, and his own heart attack. As Wolfe attempts to navigate his dual roles as a real scientist and a stage character of Gunderson's crafting, the audience bears witness to the ever-evolving collaborations of a scientist and artist, a husband and wife.*

Meredith Conti: We'd like to invite each of you to introduce yourselves. You could include land acknowledgements and pronouns, if you wish.

William DeMeritt: I'm Bill. I live in New York; the Lenape people used to live here. Hopefully some of them still do. We weren't very good to them. Well, I'm not going to say "we." I wasn't really responsible for that. I'm an actor, currently performing in *Twelfth Night* with The Classical Theatre of Harlem, and I love these two women that I get to talk to today.² It's nice to see you all.

Lauren Gunderson: Hi, Lauren Gunderson, she/her/hers. I'm in San Francisco, the unceded Ohlone land here in California. I'm a playwright and other things and a cat owner. Such a delight to be here and can't wait to chat.

Martine Kei Green-Rogers: Hi, I am Martine Kei Green-Rogers. My pronouns are she/her/hers. I am coming to you from the unceded territory of the Kickapoo, the Peoria, the Kaskaskia, the Potawatomi, and the Myaamia peoples, which carries the colonized name of the Lincoln Park neighborhood in Chicago, Illinois. And very specifically, I'm coming to you from the fifth floor of The Theatre School at DePaul University. I am super excited to be here, to have this conversation. There are so many people in this space that I totally adore.

Vivian Appler: We'd first like to ask you to speak about the connecting threads of Covid-19 and catastrophe. In media interviews during *The Catastrophist*'s premiere, Lauren, and probably also Martine and Bill, fielded questions about the play's scientific content and its sole character, Nathan, a scientist researching viruses and pandemics.³ Now, a few years into the Covid-19 pandemic, have your thoughts on Nathan's or your own catastrophizing evolved? Where do you feel we are now, culturally? How do you think the dramatized Nathan would speak about this pandemic in 2022?

WD: How would Nathan feel? It's a fucking disaster. I guess in terms of mortality and severity and maybe long-term implications, health-wise it seems to be better for most people [in 2022], but more people are getting it all the time. Fortunately, I work in an industry in which about 90 percent [of people] take it really seriously. When I did the Broadway show earlier in the year, we were on lockdown. When I do TV shows, they're locked down. With this production they're trying, but you can't always wrangle cats. I think we're paying the price for the mismanagement of the Covid virus. It's everything now, right? It's inflation. The supply chain seems worse. My mental health is fucked. I think I was doing well for a long time comparatively, but that's over. I think they're saying we're two years from an endemic, and you combine that with the fact that the United States is the worst now at handling Covid, at least in terms of industrialized nations. We are regressing culturally and sociologically and politically in a way that I saw coming. I mean, like a lot of us did, right? But it's astounding. I'm having conversations with my wife about like, can we live in this country anymore? We've got an actual virus and then we have a virus of willful ignorance.

LG: The play is about somebody who saw it coming. What he didn't see coming—and nonfictional Nathan has said this to me—the thing that was not in the model was people's resistance to common sense medicine and science, the resistance to vaccines, the resistance to the people who are legit experts, resistance to expertise. That is something that was not in the models. The difference of where we were in 2020 and now is the clarity of how frustrating it is to have a disbelief in science, which goes back several decades now of people largely in conservative parties, largely Republicans, saying science isn't real. But this is when we see the effect of it. This is when we see the final domino of that theory and the muddling of trust in science. And they're like, "oh, those nerdy scientists, don't listen to them." All of that silliness results in people dying and people literally not taking the antidote. So that was not in the play and now seems very clear. I think culturally we've learned two things. One, we now know the difference between an immunologist and an epidemiologist because Dr. [Anthony] Fauci has made sure that we understand the difference between an RNA vaccine and DNA vaccine and the difference between bacteria and virus and different kinds of viruses;[4] we now are unintentional experts in a lot of things. Perhaps not experts, but we have a new language, we have an expanded vocabulary, which is very good. But secondly, this still requires a belief that [understanding the science] is valuable, and that is not a part of the discourse in a large cross-section of American society. Of course, that hits the fan when you're sick and dying and you say, "oh, please give me the vaccine." Well, it's too late. I know I'm not the only one to have this incredible, frankly unsympathetic response, which was surprising to me. My own personal cultural confrontation is me going, "I don't care if you die, you had the chance and you didn't take it," and that is unusual for me. That's not the southern hospitality that I was raised in, you know, of love your neighbors. Honestly, the schadenfreude of it felt a little bit like, "see, I told you," which is not a healthy practice on my part, but I think that's part of the divisiveness. And it goes back to what's at the core. The facts and science are at the core [of] what this play is about: the science was there, the facts are there, the trend is there. It may not fit the narrative that you're comfortable with or that makes a good headline.

MG: My pandemic journey has traversed three very different spheres. I started Covid in New York, and then my mid-point was North Carolina, and now I'm in Chicago. Noticing how those living in these different places react/ed to the pandemic has affected the way that I look at catastrophes. And, pulling on a thread that Bill started, I'm now seeing the catastrophe of the plague itself, but also the plague of culture that came with it. And I'm going to own that in the three different spheres that I've been in, I've been horribly shocked and disappointed in varying ways as to how we as a society are thinking and dealing with this. It makes me worried, honestly, for our future, even if the pandemic does become endemic. What does this mean going forward, now that we have had this experience and this is where we are as humans?

MC: Historically, Euro-American cultures have mythologized and celebrated the solitary scientific or literary genius, typically a white man who toils away in seclusion in order to deliver his revolutionary ideas to the world. In reality, of course, the sciences and the arts are deeply collaborative disciplines, for "geniuses" like Einstein and Shakespeare as well as the rest of us. Though he is *The Catastrophist*'s only character, Nathan talks regularly of the research teams he either leads or is a part of. What does collaboration mean and look like to you? What do you hope for or potentially fear in your collaborative relationships? What was the collaborative process on *The Catastrophist* like? How do you, as theatre artists, find and sustain honest and meaningful collaborative relationships? Is there science in this?

LG: It's a fabulous question, because I think theatre and science do share that. The idea that a production's success is often misattributed to the star of the show, but it is of course writer, director, dramaturg, stage crew, designers; everyone makes the impactful thing that you see. And similarly, science has all levels and fields that come together and I'm so glad you saw this in the play. Because Nathan is the very first one to dissuade anyone of believing that he is solely responsible for any sort of discovery or result. Science has a lineage to it. It's the kind of provenance of an idea you can trace back to people and their mentors and their professors. Nathan, of course, mentions Richard Wrangham in the play, who sent him on his path to virology, really, even though Richard is a primatologist.[5] Richard said to Nathan, well it sounds like you're actually interested in *this*. And it is very similar with this play, because there are ideas in it that came in several other plays before it, some of which I mention in the preface to the published version through Bloomsbury.[6] There's a lot of this play that feels like Heidi Schreck's *What the Constitution Means to Me*.[7] The kind of confessional nature of it, the one person-ness of it. But it's not actually one person, [because] Heidi allows herself to tell her own story, and even though *The Catastrophist* is about someone else, it actually is about the writer writing it. So, I'm peeking through and that is a bold and strange choice. And I learned [this is] especially confrontative if it's a woman doing it because it hits differently than if a man is doing it. Heidi's piece suffered a similar kind of critical charge to it. And I do think there is sexism at the core of that. I also think about Rebecca Skloot's wonderful book about Henrietta Lacks where she puts herself in the story,[8] and I know Rebecca a little bit, and she said "I didn't intend to do that." I didn't either. I resisted that a long time as I tried to write about Nathan. But there is a moment when the writer is a part of the writing, and so I mention just two of those plays and works of fiction and nonfiction that allowed me to open up the idea of like, maybe it could work like this. Maybe it is OK to put me in the story because I'm part of it. Frankly, it's kind of the whole point of the story being told, where the subject and the writer are so connected that we can't be pulled apart. So anyway, the idea of collaboration in some ways starts not with the people on this call; it starts with the people in my mind and on my bookshelves and the people who allowed me to understand how to write this thing. But then of course the theatrical collaboration [for *The Catastrophist*] was particular because of the time in which we were doing it. The choice to write a one-person play was very strategic and necessary because this was pre-vaccines. We were making a play when we should not be making plays, and so the question was how can we write it so it can work safely and still feel like theatre? So, we're not Zooming it, we're not making it an audio play . . . I mean, great audio theatre feels very much like theatre. But this idea of filming it in the theatre, with theatrical lighting, and on a stage, was different than what we were seeing around. It required a ton of collaboration, largely on Marin [Theatre Company]'s part. Martine and I were never in the same space together. I was never in the same space with Bill except for a random, masked hike; we both had said to each other, "I would like to be near you in human life." When they were actually producing *The Catastrophist*, it was Jasson Minadakis, the director, and Peter Ruocco, our cinematographer, and Bill largely making all the decisions in the room. I was on Zoom the whole time and we were texting constantly and I was editing from afar in terms of, well, can we change this line? Can you skip this part? Can you change this? We were doing that collaborative work, but we had to find different ways to do it.

MG: Thank you for that, Lauren. In thinking about these questions, "how do you as theatre artists find and sustain productive, honest, and meaningful collaborative conversations? Is there science in this?"—I think the answer is yes. *The Catastrophist* was such a strange collaboration because it was atypical of what normally happens in the theatre. Part of our collaborative process was figuring out how to make this work. However, most of us came into this process with an advantage, which is that in some way, shape, or form we were already connected to someone in the virtual space of the show. There were past collaborations and established paths of friendship between us all. And to be honest, a lesson that I hope we learned from the pandemic is that the regional theatre model has gotten away from sustained collaboration. You come in for this project, and then you move on to a different project. Because of the pandemic, if we were creating anything, we were doing it in these kinds of [remote] spaces where we had to lean on prior relationships, but we also brought new collaborators into the fold and expanded our circles. I think the science to that is how do we find our people and then how do we continue to find ourselves in virtual and real spaces with those people?

WD: I consider myself a very collaborative person. Maybe to a fault, I don't know. So doing this show and not being able to physically be in a room with anyone was incredibly difficult. That being said, Lauren, Martine, Jasson, Nathan, and everyone involved were really wonderful about doing as much as could be done safely. This might still be the case, but certainly at the time when the show was released it was the case, that we were among the best examples of how to do accessible, streamed, meaningful theatre. Leaving the subject matter and the quality of the play aside, which I think are of the utmost importance, but just in terms of the production value and keeping costs down for tickets, I think all that was great. It was great to innovate new ways to collaborate, because it's not about how hard you collaborate, right, it's about how well you collaborate. It's not about the quantity, it's about the heart. The pandemic forced us all to work differently. We had this one rehearsal, Lauren's probably sick of me talking about it, but where I'm doing a table read alone in the Zoom. I'm reading the script off of a shared Google Doc and Lauren is literally changing the lines as I'm reading. And I had to say to her, "OK, is this because I suck, or because you had an idea?"

LG: It was not because you suck!

WD: I didn't know Jasson at the time, but Martine and Lauren are in that rare echelon of collaborators in that they're very generous. They're as intelligent as they are generous, right? Ego never enters the room for them.

VA: *The Catastrophist* premiered as a play to be streamed due to the circumstances of Covid-19, and there are some lovely invitations for audience interaction, particularly with regard to smell. This is a multipart question: To what extent was audience participation imagined as an integral part of the play? How did you consider audience presence when performing on the Marin Theatre stage by yourself, Bill, and for Martine and Lauren, how do these suggested audience interactions work dramaturgically as embodied moments?

WD: When I first read the script and there was all this written about the smells and stuff, I was really looking forward to that. It is one of those things that in execution might be tougher to pull off, but should still be attempted at some point. And as a theatre actor, you're always aware of the audience, even if you shouldn't be aware of the audience, you know? But then there are these breaking-the-fourth-wall moments. What Jasson and Lauren invited me to do was a combination. The show opens and I'm talking to the camera. Most of the time I'm talking to the camera. I had to start thinking of the camera as . . . I had to give it a role. I had to give the lens a role that was an invested role, not just as the spectator. That can change throughout, too, and certainly it did in parts of the play where I'm talking about Nathan's children, my children. There were a couple of moments where we located people in the audience, Lauren. And then there were parts where I was encouraged to think about it as a TedTalk. So, I was always in conversation in *The Catastrophist*, but the person I was in conversation with changed; their location and their physical and emotional proximity changed.

MG: Dramaturgically, I think one of the things that's always so interesting to me about smell is that smell is visceral. It's so visceral! To this day, I cannot smell Polo Blue and not think about an ex-boyfriend. Inviting the audience to smell, even if that smell isn't there, asks our brains to conjure something. Smell is attached to emotion, in some way, shape, or form.[9] The world that Lauren created is very intimate and [the evocations of smells] make it even more intimate. I think you find yourself trusting your main character a little bit more because they're evoking smells that you may or may not know. It really sort of sucks you into the world. Especially considering we have a piece of theatre that was being done on a screen, I think it was a brilliant thing dramaturgically to try and bring us back into an intimate space.

LG: Yeah, *The Catastrophist* was designed to make an at-home viewing extrasensory and special. And in fact, we were thinking up things that you couldn't do in the theatre really well, right? So that was part of the impetus of that. And we kept it in the script even though we didn't quite execute it in production. The offerings are still there, though, and we even thought for a while, like, could we send people little packets of coffee grounds and beeswax? I'm like, "well we're in a pandemic, so maybe don't send people things to smell." But we did try to include smells that were in most people's homes. But I do think there's power, as Martine said, in even mentioning a smell. It's so tied to memory, and it combines the character's memory with whatever your memory is. That's that magic of theatre, that blending. This goes back to your question about collaboration; we didn't actually mention the audience, right? And for this play in particular, that is a big, big reach and the final handshake is with an audience that we can't see. We can't adjust based on their feelings like we usually do in the theatre. We can't say "oh, God, the jokes aren't landing." We didn't know. But in this way we know, just psychosomatically, that an audience will have something to experience when you say coffee and/or beeswax candles. Through the screen, something approximating a community was hopefully feeling something at the same time.

MC: The contributions to this volume, which is the second in a book series entitled *Identity, Culture, and the Science Performance*, are centered around scientific or performance phenomena that are hidden or inscrutable or incredibly distant, but that nevertheless impact our daily lives. In *The Catastrophist*, latent threats to personal health, such as congenital heart conditions and cancer, are discussed alongside the enormous questions pertaining to microscopic viral and cellular life. How do you all envision *The Catastrophist*'s participation within public health conversations or initiatives that raise awareness about hidden or misunderstood health dangers? Lauren, you have already connected this play to contemporary Americans' trust and/or mistrust in science. Did this play educate? Was that its intention? What was its role within public health discourse?

LG: I love this theme of hiddenness, because the way I write and structure drama is . . . a lot of it is hidden. In terms of the public health of it, yes, there is work in talking about viruses, talking about statistics, talking about risk. There's the statistics and probability of trying to get people to understand what is actually dangerous versus what appears dangerous. And that is a lot of the confusion around public health. But one thing I think is most critical about the play in terms of adding value to public trust in science is seeing—and Bill's performance amplifies and deepens this so much—the thrill of science. You're watching a scientist who is in it to win it and loves his field, as opposed to making science boring and dowdy and difficult. A long, long time ago Nathan was called like "the Indiana Jones of Viruses," which he hates, but I adore, of course. But that character made people fall in love with archeology. I was all in for dinosaurs and archaeology because of *Jurassic Park* and *Indiana Jones* and seeing scientists as swashbuckling and passionate; it honestly helps people want to do science! This is what we need, not just to trust ones we have but to make the next generation [of scientists].

Now, in terms of "hiddenness," the play actually has so many levels of hidden. The whole structure is actually hidden until the very end, when we realized through a reveal, the kind of tumbling reveal at the end of where he actually is, what he's actually waiting for, and why he's in this space at all. Another level is the audience: we show empty seats. The audience is hidden, is behind screens and thousands of miles away in different countries; we know they're there, but it's a very different experience. It's even written into the show. Our character sees that he's in a theatre and yet there is no audience. That means something, and at the beginning he doesn't quite know what that means, but as the play goes on, we start to understand what it means. What Nathan studies is also hidden, right? The unseen, the unseeable. The character of the father is so important and so *there*, but *not* there and that's part of the heartbreak. His wife is there telling Nathan things, but we don't see her until just a glimpse at the very end. There's so much that this character has to trust despite them being hidden; that leans over into Judaism, the religion of it, to where we are trusting something we can't see. There's so much to the play that deals with what is invisible but powerful.

MG: The dramaturg in me is always looking for "what is our contemporary relevance?" And obviously we're in the middle of a pandemic, so you can't get more contemporarily relevant than that. When we were working on *The Catastrophist*, I was still actively teaching. I told my dramaturgy students about this project and about our collaborative process working through Zoom, and some of the students found themselves innately curious enough to go find Nathan's book in the library.[10] And then they read it together, essentially assigning different chapters to different people to read. And then they came together and had a conversation. So, me just telling them about that process made them want to educate themselves a bit about the play's themes. In terms of thinking about public health initiatives and raising awareness, that is what we're meant to do. We are meant to ask probing questions about the universe around us, and hopefully some part of the story we tell resonates enough to make someone say I need to know more. Maybe the personalization [in *The Catastrophist*] will cause someone to go do something—get something checked out about their health that they knew about from their family history. I have grandiose hopes for the idea. Yes, I know that not everyone who encounters *The Catastrophist* will get themselves checked out for potential health ailments. But just maybe, they might remember that time that they saw this play or read this play and think about how could they do something a little differently for their well-being.

VA: Our last big question centers on identity as it intersects with science. Much of the appeal of Lauren's science dramas has to do with her capacity to imagine very specific personal circumstances for characters whose stories have been abstracted or hidden from history, oftentimes due to gender, racial, ethnic, or social discrimination and inequities. This play, which represents a working scientist onstage, is deeply personal for Lauren and her husband. As a playwright, dramaturg, or actor, how did you approach identity and representation in *The Catastrophist*, particularly in how the play depicts scientists and scientific communities to audiences?

WD: Lauren told me at some point that it was very important that the actor portraying Nathan be Jewish. As a Black Jew, as someone of mixed ethnicity, I rarely get to play any of those things. Onstage I've been—for lack of a better word—"allowed" to play Black characters since about 2016. In grad school I never did any of the Black shows, and on television I've only ever played a specifically Black character once. Outside of *The Catastrophist*, I've never played a Jewish person on film, and in 2019 at the Oregon Shakespeare Festival I was in Paula Vogel's *Indecent* as my first time being a Jewish character.[11] It was very touching and meaningful to me that Lauren wanted me to do the show, both as a Jew and as a Jew of color, because this country is still very skewed in its definition of what Judaism means, or even what Ashkenazi Judaism means. And when people of color are playing scientists on film and in television, it's usually a certain type of scientist. It's never the Indiana Jones or Laura Croft Tomb Raider type of scientists, right? It's BD Wong or Sam Jackson in *Jurassic Park*. And they're great, I love those guys, but they're always side characters who report to someone. But *The Catastrophist*'s main character, the only character, is a scientist and a little bit of the Indiana Jones model. We don't necessarily see Nathan jumping and running, but I'm in the jungle and there are animals and it's dangerous and adventurous . . . there are bandits and I'm fighting against the government. And the character is Jewish and the show interacts with his Judaism a lot, but it's not a show about Judaism. When we do shows that are, like, Jewish shows it's always about the Holocaust, or immigration, or, some biblical story. It's *Prince of Egypt* or *Schindler's List* or *An American Tale*. It's great to work on a story where Judaism is important, but it's not about some . . . generational trauma.

LG: I mean, what an interesting question. Well, there's my own identity as a woman telling a story of science, and that already gets spicy. And a woman telling a man's story; that becomes another level of overreach in some people's minds. And then the third level of telling, I added myself to that story, writing as a wife and mother, which provides another level of confrontation to some people. It is so hard for people to let this play be what it is without trying to tear it apart, because having a wife tell a husband's story brings out all the sexism. This is a person of note, a scientist, and the play depicts a kind of camaraderie between Nathan and me that I think is still oddly and unfortunately strange. Like, "how would you let your wife say that? How would you let your wife write about that?" There are reactions as though I have commandeered his life and legacy. It's an interesting part of sexism I didn't know was going to be a part of this whole process. In retrospect, I'm like "God, that is infuriating." But at the same time, everything in the play needs to be there this way, and it is a massive credit to Nathan that he agreed to do this bonkers project, to do something so vulnerable, and to trust me as a dramatist telling his own story. It makes people examine what they think of wives and husbands and what they think about a wife's, a mother's, a woman's ability to enter a "male" arena. I'm incredibly proud about that part of it because of not just what it says about me, but mostly what it says about Nathan and about us as a team. And there was no question among our collaborators . . . no one was ever like "Lauren, should you?" Like, of course not. The utter confidence that that this team had in me to do this hard thing!

VA: When you move into telling the story of the heart attack [in the play], he gets frustrated with you, and starts yelling at you directly.

LG: This is the idea of what science knows versus what art knows. That moment specifically, that's probably the most fictionalized part. Nathan loves theatre and truly believes in the human truth and necessity of it. But the idea that his character is willing to say theatre is useless and science is necessary. I think I tried to speak the quiet part out loud by having Nathan say "you think you're going to save me with a play?" Which is honestly what I sometimes tell myself when I'm writing, like "God, what are you doing? You're going change the world with fiction? Come on." But of course, it does, in strange and unexpected ways. It allows conversations to happen that wouldn't happen, and conversations are how humans engage and prepare to do the work.

The stories of scientists have been done in less personal ways than this play. It's a very important part of what I do, partly because as a young writer I knew I wanted to write theatre about the most important things on Earth. I didn't understand why anyone would not do that. Like, if you're going to do this craft you've got to write about the important stuff. And when I turned to the biggest things that I could think of, it was the stories of scientists and the stories of discoverers and explorers and people who changed the way we see the world and each other. In every play about scientists that I've written, there is a lot of personal backstory and consequence, emotional heart stuff, family stuff. Nothing like *The Catastrophist*, though, because it really is so bare and raw in terms of what else is going on in this human life. A scientist is not just a brain. A scientist is not just a series of experiments they run. They're human beings that are fraught and scared and dealing with a ton of shit that you can't science your way out of. You can't science your way out of losing your father. You have to sit with the reality of it. You can't change it. And so that sense of, in some ways, this play pits the scientist up against all the stuff they cannot do anything about with their skill set, and that's part of why the play has such tension. And all of my plays push the characters to the point where they are past their capability. That's what I think plays are for, is really testing humans in their resilience to see how they solve stuff, how they survive things, if they survive things. That's where I have to go as a dramatist, to take this person—in this case a person I know and love—and take them to the place that is beyond their capability to handle, which was a lot to ask my husband to let me do. This is a play representing a real person, but it's that liminal space, the "is it real or isn't it?" of the theatre, right? I think that dance is what we were testing the entire time, over and over. And Bill and Nathan look identical as children, we realized from pictures of them.

WD: I don't understand it.

LG: It's so weird.

WD: I genuinely am not mad about it. I'm just deeply confused.

LG: It's kind of spooky. Not [to] speak for you, Bill, but some of your experiences in life mirror some of Nathan's in a way that was also kind of—

WD: That was also surprising. The loss of a parent and yeah, that type of stuff.

LG: Yeah, and wrestling with religion, too. We found ourselves having those ideas from real life spill into this play, which you always do in the theatre, but in this case there was a lot of extra electricity in between those things.

MG: I think for me, my job was to live in between both worlds that Bill and Lauren talk about. The thing that I was keenly aware of was how we're depicting scientists and scientific communities and how audiences would receive these representations, but [it] was also my job to just trust Lauren because she's writing about her partner. I don't know if I could say that, dramaturgically, anyone would know her partner better than her. And I consider Bill a really good, close friend, and his perspective on the character he is creating meant a whole lot to me, too, in terms of thinking through identity and representation.

MC: We'd love to conclude with any final thoughts you have on your work with *The Catastrophist*. We know this is a reunion of sorts for you all!

WD: I miss it. You know, it was a wonderful thing to create. No matter what happens, I think it's going to be one of the best things I have done. Lauren mentioned sexism in the way the piece was received and without going off on a whole tangent, I think this show could have had a larger life if it wasn't for Jesse Green and the *New York Times*. I don't know if you all read that review. I think he said I was intelligent and sexy and that was nice, and the rest of it seemed to be an opportunity for him to just assassinate Lauren's career versus this one show. I know everyone's trying to be George Bernard Shaw, but maybe review it with a little bit more grace . . . At any rate, I started thinking about theatre criticism in a way that I hadn't really before, because I generally don't pay attention to it. I hope the play continues on, whether it's through people studying it, or we do it again, or Netflix buys it. I think regardless it's going to be a hell of an interesting time capsule.

MG: Yeah, I will say that it was definitely a transformational period in my experience of thinking about what theatre can do, and the resilience of theatre in the face of a pandemic, and lastly, how storytelling is always going to exist no matter what. Flat out, no matter what, storytelling will always exist. I feel very blessed that I got a chance to participate in this project. I do hope that it gets to have a new life that is free of the bias of reviewers.

LG: It was such a rarified time when we created this, and I think it's worth noticing the bravery of doing anything, any theatre, but especially this kind of thing. I will always look back and feel a kind of rush of creativity and collaboration. In my twenty years in the theatre, this was probably the most special [experience] because of what we were up against and what we created. It was a landmark streamed American theatre production that we hadn't seen before, and we didn't know entirely what we were doing when we did it. It was like building the boat as you're sailing it, you know. But I think that is the theatre at its best. Theatre can be really responsive; it actually can be quicker than TV and film. It doesn't have to be a three- to ten-year process. It can be months. I'm incredibly grateful for Bill and Martine and the folks at Marin and Jasson and of course Nathan, who had all of the cost and none of the control. I'm so proud of the thing we made.

Notes

1. For further information on our panelists, please see williamdemeritt.com, laurengunderson.com, and martinekeigreenrogers.com.
2. The *New York Times* named The Classical Theatre of Harlem's *Twelfth Night* a "Critic's Choice." Laura Collins-Hughes, "'Twelfth Night' Review: A Shot of Joy under a Darkening Sky," *New York Times*, 12 July 2022.
3. The premiere of *The Catastrophist*, coproduced by the Marin Theatre Company and Round House Theatre, streamed January 26–July 25, 2021.
4. In a December 2020 CNBC Healthy Returns Livestream, Fauci dispelled myths about Covid vaccines. Meg Tirrell, "Dr. Anthony Fauci with Meg Tirrell for special edition of Healthy Returns," *CNBC*, 16 December 2020, livestream, https://www.cnbc.com/2020/12/16/cnbc-transcript-dr-anthony-fauci-speaks-with-cnbcs-meg-tirrell-live-during-the-cnbc-healthy-returns-livestream-today.html.
5. Wrangham is the founder of the Kibale Chimpanzee Project and Ruth B. Moore Professor of Biological Anthropology at Harvard University. For more about the Kibale Chimpanzee Project, see Kibale Chimpanzee Project: Research, conservation, and education, kibalechimpanzees.wordpress.com.
6. Lauren Gunderson, *The Catastrophist* (London: Methuen Drama, 2021).
7. Heidi Schreck, *What the Constitution Means to Me* (New York: Samuel French, 2017).
8. Rebecca Skloot, *The Immortal Life of Henrietta Lacks* (New York: Random House, 2010).
9. In the first volume of this book series, Meredith Conti contends that smells are dramaturgical tools that critically shape performance. See "Ether, Sawdust, Sweat, and Blood: Of Smells and Victorian Operating Theatres," in *Identity, Culture, and the Science Performance, Volume 1: From the Lab to the Streets*, ed. Vivian Appler and Meredith Conti (London: Bloomsbury Methuen, 2022), 141–58.
10. Nathan Wolfe, *The Viral Storm: The Dawn of a New Pandemic Age* (New York: Times Books, 2011).
11. Vogel's *Indecent*, directed by Shana Cooper for the Oregon Shakespeare Festival, ran from July 4 to October 27, 2019, in the Angus Bowmer Theatre.

Creative Interlude

From *The Catastrophist*

Lauren Gunderson

Scene 3: Beeswax

"Beeswax" is visible to the audience and to him somewhere.

Sniffs again.

Beeswax.

Audience smells: beeswax.

Smell is the most powerful sense for most mammals.
Evocative sense.
Smell is a conjurer.
It yanks us around in time, doesn't it?

Smells again. The smell pulls the memories back to him.

Childhood.
Home.
Synagogue.
Service.
Shiva.
Candles lit.
In someone's honor.

This smell usually meant death to me.

There was a lot of it in my family.

Starting to put things together.

One data point to notice right off the bat is that the men in my family don't tend to live past forty. My grandfather, my uncle Itzy, my cousin Jeff, my other cousin, all gone by forty. Heart disease. All of them.

Candles lit in their honor.

The smell makes him think . . .

My father survived his massive heart attack at thirty-nine but just barely.
I was ten.

He was in the hospital for weeks. I was told he would die. They told me I wasn't allowed to see him either, I couldn't go in, I was a kid and kids aren't allowed and it made absolutely no critical sense why that was the case, and I was so furious at every single person in that hospital because everyone was just following the rules when the rules were idiotic.

"You're going to stop me from seeing my own father, from touching him one last time before he dies in that room right there? He's right there and you won't let me in there to say goodbye?!"

I decided I was never going to be a doctor. No. No way. It turns out I am a doctor. PhD in Virology. Specifically, "Doctor of Science in Immunology and Infectious Diseases." Harvard. My grandmother would like me to make sure and mention the Harvard part.
But even Nana used to say that the PhD is just until I become a real doctor.

A joke, but in my family a joke is proof of concept.

This is my Seventies Jewish upbringing, my parents raised us to do good, be a good person, always act in the effort of making the world better. Tikkun Olam is the Jewish concept that means: "to heal the world." That was important to my dad. Anyone can help to heal the world using the talents they've been blessed with. What will you do?

In my home on a shady curved street in West Bloomfield, Michigan, Judaism was a massive part of my life—we kept Kosher for most of my childhood, my dad was a Jewish communal servant, the director of a Jewish home for the elderly, we went to Jewish camp for months in the summer—that's where my mom and dad met actually—in the woods singing camp songs!—anyway these Jewish ideas of social justice and do-gooding were definitive in my family. My mom's family was reform, but my dad was raised orthodox, and Judaism was a part of everything we did.

The other concept that was taught to me was this from the Talmud; "whoever saves one life saves the entire world."

Maybe I can do better than one.

SELECTED BIBLIOGRAPHY

Adams, Rachel, Benjamin Reiss, and David Serlin, eds. *Keywords for Disability Studies*. New York: New York University Press, 2015.
American Psychiatric Association, ed. *Diagnostic and Statistical Manual of Mental Disorders: DSM-5*. Arlington: American Psychiatric Association, 2013.
Appler, Vivian and Meredith Conti, eds. *Identity, Culture, and the Science Performance, Volume 1: From the Lab to the Streets*. London: Bloomsbury Methuen, 2022.
Arons, Wendy and Theresa J. May, eds. *Readings in Performance and Ecology*. New York: Palgrave Macmillan, 2012.
Bahr, Ehrhard. *Weimar on the Pacific: German Exile Culture in Los Angeles and the Crisis of Modernism*. Berkeley: University of California Press, 2007.
Ball, Philip. *Serving the Third Reich: The Struggle for the Soul of Physics under Hitler*. Chicago: University of Chicago Press, 2014.
Barry, Andrew and Georgina Born, eds. *Interdisciplinarity: Reconfigurations of the Social and Natural Sciences*. London: Routledge, 2013.
Bennett, Michael Y. *Narrating the Past through Theatre: Four Crucial Texts*. New York: Palgrave, 2013.
Bentley, Joanne. *Hallie Flanagan: A Life in the American Theatre*. New York: Alfred A. Knopf, 1988.
Bohr, Neils. *Atomic Physics and Human Knowledge*. 1949. New York: Science Editions, 1961.
Bormann, Natalie and Michael Sheehan, eds. *Securing Outer Space: International Relations Theory and the Politics of Outer Space*. Abingdon: Routledge, 2009.
Born, Max. *The Born-Einstein Letters: Friendship, Politics and Physics in Uncertain Times*. Trans. Irene Born. New York: Macmillan, 1971.
Born, Max. *Die Relativitätstheorie Einsteins und inren physikalischen Grundlagen*. 1st ed. Berlin: Springer-Verlag, 1920.

Borusso, Marinella Miano. *Hombre, Mujer y Muxe en el Istmo de Tehuantepec*. Mexico City: CONACULTA-INAH, 2002.

Burelle, Julie. *Encounters on Contested Lands: Indigenous Performances of Sovereignty and Nationhood in Québec*. Evanston: Northwestern University Press, 2019.

Butler, Judith. "Performative Acts and Gender Constitution: An Essay in Phenomenology and Feminist Theory," *Theatre Journal* 40, no. 4 (1988): 35–49. https://doi.org/10.2307/3207893.

Butler, Judith. *Undoing Gender*. New York: Routledge, 2015.

Cañizares-Esquerra, Jorge. *Nature, Empire, and Nation: Explorations of the History of Science in the Iberian World*. Stanford: Stanford University Press, 2006.

Canning, Charlotte M. *On the Performance Front: US Theatre and Internationalism*. Houndsmills, Basingstoke, Hampshire: Palgrave Macmillan, 2015.

Carson, Cathryn. *Heisenberg and the Atomic Age: Science and the Public Sphere*. Cambridge: Cambridge University Press, 2010.

Chaudhuri, Una and Holly Hughes, eds. *Animal Acts: Performing Species Today*. Ann Arbor: University of Michigan Press, 2014.

Chiari, Sophie. *Shakespeare's Representation of Weather, Climate and Environment*. Edinburgh: Edinburgh University Press, 2019.

Cohen, Bruce M. Z., ed. *Routledge International Handbook of Critical Mental Health*. London: Routledge, 2019.

Confino, Alon. *A World without Jews: The Nazi Imagination from Persecution to Genocide*. New Haven: Yale University Press, 2014.

Conti, Meredith. *Playing Sick: Performances of Illness in the Age of Victorian Medicine*. London: Routledge, 2018.

Cooper, Melinda and Catherine Waldby. *Clinical Labor: Tissue Donors and Research Subjects in the Global Bioeconomy*. Durham: Duke University Press, 2014.

Curtis, Harriet and Martin Hargreaves, eds. *Kira O'Reilly: Untitled (Bodies)*. London: Intellect Books and Live Art Development Agency, 2018.

Day, Stuart A., ed. *Performances that Change the Americas*. Abingdon, Oxon: Routledge, 2022.

Decker, Hannah S. *The Making of DSM-III: A Diagnostic Manual's Conquest of American Psychiatry*. Athens, OH: Ohio University Press, 2013.

Dickens, Peter and James S. Ormord. *The Cosmic Society: Towards a Sociology of the Universe*. London: Routledge, 2007.

Dickens, Peter and James S. Ormrod, eds. *The Palgrave Handbook of Society, Culture and Outer Space*. Houndsmills, Basingstoke, Hampshire: Palgrave, 2016.

Donger, Simon, Simon Shepherd and ORLAN. *ORLAN: A Hybrid Body of Artworks*. London: Routledge, 2010.

Drake, Frank and Dava Sobel. *Is Anyone Out There? The Scientific Search for Extraterrestrial Intelligence*. New York: Delacorte Press, 1992.

Flanagan, Hallie. *Arena: The Story of the Federal Theatre*. 1940. New York: Limelight Editions, 1985.

Fraden, Rena. *Blueprints for a Black Federal Theatre, 1935–1939*. Cambridge: Cambridge University Press, 1994.

Franzmann, Vivienne. *Bodies*. London: Nick Hern Books, 2017.

Geiger, Hans, Karl Freidrich Franz, and Christian Scheel. *Handbuch der Physik*. Berlin: Julius Springer, 1928.

Ghostkeeper, Elmer. *Spirit Gifting: The Concept of Spiritual Exchange*. Nanaimo: Strong Nations Publishing, 2007.

Goffman, Erving. *The Presentation of Self in Everyday Life*. New York: Anchor Books, 1959.

Grehan, Helena and Peter Eckersall, eds. *The Routledge Companion to Theatre and Politics*. London: Routledge, 2019.

Griffin, Gabriele and Doris Leibetseder, eds. *Bodily Interventions and Intimate Labour: Understanding Bioprecarity*. Manchester: Manchester University Press, 2020.

Gurr, Andrew and Farah Karim-Cooper, eds. *Moving Shakespeare Indoors: Performance and Repertoire in the Jacobean Playhouse*. Cambridge: Cambridge University Press, 2014.

Harding, Sandra. *Objectivity and Diversity: Another Logic of Scientific Research*. Chicago: The University of Chicago Press, 2015.

Harding, Sandra. *Sciences from Below: Feminisms, Postcolonialities, and Modernities*. Durham: Duke University Press, 2008.

Harpin, Anna and Juliet Foster, eds. *Performance, Madness and Psychiatry: Isolated Acts*. New York: Palgrave Macmillan, 2014.

Harrison, Laura. *Brown Bodies, White Babies: The Politics of Cross-Racial Surrogacy*. New York: New York University Press, 2016.

Hayles, N. Katherine. *Chaos Bound: Orderly Disorder in Contemporary Literature and Science*. Ithaca: Cornell University Press, 1990.

Heisenberg, Werner. *Die Physikalischen Prinzipien der Quantentheorie*. Liepzig: Von S. Hirzel, 1930.

Hippe, Christian, ed. *Bild und Bildkünste Bei Brecht, Brecht Tage*. Berlin: Matthes & Seitz, 2011.

Hunter, Lynette. *Politics of Practice: A Rhetoric of Performativity*. London: Palgrave Macmillan, 2019.

Jackson-Schebetta, Lisa. *Traveler, There Is No Road: Theatre, the Spanish Civil War, and the Decolonial Imagination in the Americas*. Iowa City: The University of Iowa Press, 2017.

Jones, James H. *Bad Blood: The Tuskegee Syphilis Experiment*. New York: The Free Press, 1981.
Joyce, Arthur A. *Mixtecs, Zapotecs, and Chatinos: Ancient Peoples of Southern Mexico*. Oxford: Blackwell Publishing, 2009.
Kaur Chohan, Satinder. *Made in India*. London: Samuel French, 2016.
Kojevnikov, Alexei, Helmuth Trischler, and Cathrun Carson, eds. *Weimar Culture and Quantum Mechanics: Selected Papers by Paul Forman and Contemporary Perspectives on the Forman Thesis*. London: Imperial College, 2011.
Magelssen, Scott. *Performing Flight: From the Barnstormers to Space Tourism*. Ann Arbor: University of Michigan Press, 2020.
Magli, Giulio. *Archaeoastronomy: Introduction to the Science of Stars and Stones*. Cham: Springer, 2016.
Marche, Jordan. *Theaters of Space and Time: American Planetaria, 1930–1970*. New Brunswick: Rutgers University Press, 2005.
Marshall, Jonathan W. *Performing Neurology: The Dramaturgy of Dr. Jean-Martin Charcot*. New York: Palgrave Macmillan, 2016.
McCook, Stuart. *States of Nature: Science, Agriculture, and Environment in the Spanish Caribbean, 1760–1940*. Austin: University of Texas Press, 2002.
McDermott, John F. and Naleen Naupaka Andrade, eds. *People and Cultures of Hawai'i: The Evolution of Culture and Ethnicity*. Honululu: University of Hawai'i Press, 2011.
Mercelis, Joris. *Beyond Bakelite: Leo Baekeland and the Business of Science and Invention*. Cambridge, MA: MIT Press, 2020.
Mermikides, Alex and Gianna Bouchard, eds. *Performance and the Medical Body*. London: Bloomsbury Methuen Drama, 2016.
Mignolo, Walter D. *Local Histories/Global Designs: Coloniality, Subaltern Knowledges, and Border Thinking*. Princeton: Princeton University Press, 2000.
Miller, Paul. "The Hidden Code," http://djspooky.com/the-hidden-code/.
Miller, Paul (DJ Spooky That Subliminal Kid). *Rhythm Science*. Cambridge, MA: MIT Press, 2004.
Monge, José Trias. *Puerto Rico: The Trials of the Oldest Colony in the World*. New Haven: Yale University Press, 1997.
Nelson, Cary and Lawrence Grossberg, eds. *Marxism and the Interpretation of Culture*. Chicago: University of Illinois Press, 1988.
Nelson, Robin. *Practice as Research in the Arts (and Beyond): Principles, Processes, Contexts, Achievements*. 2nd edition. Cham: Palgrave Macmillan, 2022.
Nietzsche, Friedrich. *Die Geburt die Tragödie*. Berlin: Walter de Gruyter, 1967.
Nietzsche, Friedrich. *Nietzsche contra Wagner: Aktenstücke eines Psychologen*. Leipzig: C.G. Neumann, 1889.

O'Shea, Janet. *At Home in the World: Bharata Natyam on the Global Stage*. Middletown: Wesleyan University Press, 2007.
Osborne, Elizabeth A. *Staging the People: Community and Identity in the Federal Theatre Project*. New York: Palgrave Macmillan, 2011.
Panksepp, Jaak and Lucy Biven. *The Archaeology of Mind: Neuroevolutionary Origins of Human Emotions*. New York: W.W. Norton & Company, 2012.
Parker, Stephen. *Bertolt Brecht: A Literary Life*. London: Bloomsbury, 2014.
Pérez Jr., Louis A. *The War of 1898: The United States and Cuba in History and Historiography*. Chapel Hill: University of North Carolina Press, 1998.
Pick, Zuzana M. *Constructing the Image of the Mexican Revolution: Cinema and the Archive*. Austin: University of Texas, 2010.
Planck, Max. *Eight Lectures on Theoretical Physics*. Trans. A. P. Willis. New York: Columbia University Press, 1915.
Portilla, Miguel Leon. *La Filosofía Náhuatl*. Mexico City: National Autonomous University of Mexico Press, 2016.
Pratt, Mary Louise. *Imperial Eyes: Travel Writing and Transculturation*. New York: Routledge, 2008.
Quinn, Susan. *Furious Improvisation: How the WPA and a Cast of Thousands Made High Art out of Desperate Times*. New York: Walker & Company, 2008.
Rancière, Jacques. *The Politics of Aesthetics: The Distribution of the Sensible*. New York: Continuum, 2004.
Said, Edward W. *Culture and Imperialism*. New York: Alfred Knopp, 1993.
Said, Edward W. *Orientalism*. New York: Pantheon Books, 1978.
Schwartz, Bonnie Nelson. *Voices from the Federal Theatre*. Madison: University of Wisconsin Press, 2003.
Sedgwick, Mark, ed. *Key Thinkers of the Radical Right: Behind the New Threat to Liberal Democracy*. Oxford: Oxford University Press, 2019.
Serres, Michel. *Atlas*. Paris: Editions Julliard, 1994
Serres, Michel. *The Parasite*. Trans. L. R. Schehr. Minneapolis: University of Minnesota Press, 2007.
Serres, Michel, with Bruno Latour. *Conversations on Science, Culture, and Time*. Ann Arbor: University of Michigan Press, 1995.
Shakespeare, William. *The Tragedy of King Lear*. 1606. Folger edn. Eds. Barbara A. Mowat and Paul Werstine. New York: Simon and Shuster, 2015.
Shapin, Steven. *Never Pure: Historical Studies of Science as if it was Produced by People with Bodies, Situated in Time, Space, Culture, and Society, Struggling for Credibility and Authority*. Baltimore: The Johns Hopkins University Press, 2010.

Shepherd-Barr, Kirsten, ed. *The Cambridge Companion to Theatre and Science*. London: Cambridge, 2020.
Snow, C. P. *The Two Cultures and the Scientific Revolution*. Cambridge: Cambridge University Press, 1959.
Sofer, Andrew. *Dark Matter: Invisibility in Drama, Theater, and Performance*. Ann Arbor: University of Michigan Press, 2013.
Spengler, Oswald. *Der Untergang Des Abendlandes*. 1st ed. München: C.H. Beck'sche, 1919.
Stern, Steve J. *The Secret Story of Gender: Women, Men & Power in Late Colonial Mexico*. Chapel Hill: The University of North Carolina Press, 1995.
Sturgess, Keith. *Jacobean Private Theatre*. London: Routledge & Kegan Paul, 1987.
Taylor, Diana and Sarah J. Townsend, eds. *Stages of Conflict: A Critical Anthology of Latin American Theater and Performance*. Ann Arbor: University of Michigan Press, 2008.
Thompson, L. *Nuestra Isla y su Gente: La Construcción del "Otro" Puertorriqueño en "Our Islands and Their People."* San Juan: Centro de Investigaciones Sociales, 2007.
Tobias, Philip V., Michael A. Raath, Jacopo Moggi-Cecchi and Gerald A. Doyle, eds. *Humanity from African Naissance to Coming Millennia*. Florence: Firenze University Press, 2001.
Tuck, Eve and K. Wayne Yang. "Decolonization is Not a Metaphor." *Decolonization: Indigeneity, Education & Society* 1, no. 1 (2012): 1–40.
Turner, Henry S., ed. *Early Modern Theatricality*. Oxford: Oxford University Press, 2013.
United Nations. *United Nations Treaties and Principles on Outer Space: Text of Treaties and Principles Governing the Activities of States in the Explication and Use of Outer Space*. 1976. New York: United Nations, 2002, https://www.unoosa.org/oosa/en/ourwork/spacelaw/treaties/outerspacetreaty.html.
Vakoch, Douglas A. *Archaeology, Anthropology, and Interstellar Communication*. Washington, D.C.: NASA, 2014.
Waldby, Catherine. *The Visible Human Project: Informatic Bodies and Posthuman Medicine*. London and New York: Routledge, 2000.
Walker, Nick. *Neuroqueer Heresies: Notes on the Neurodiversity Paradigm, Autistic Empowerment, and Postnormal Possibilities*. Fort Worth: Autonomous Press, 2021.
Washuta, Elissa and Theresa Warburton, eds. *Shapes of Native Nonfiction: Collected Essays by Contemporary Writers*. Seattle: University of Washington Press, 2019.

Witham, Barry B. *The Federal Theatre Project: A Case Study*. Cambridge: Cambridge University Press, 2003.

Woolford, Andrew, Jeff Benvenuto and Alexander Laban Hinton. *Colonial Genocide in Indigenous North America*. Durham: Duke University Press, 2014.

CONTRIBUTORS

Vivian Appler is Kay Parker Professor of Theatre Arts and Associate Professor of Performance Studies at the University of Georgia, USA. She is coeditor, with Meredith Conti, of the two-volume collection *Identity, Culture, and the Science Performance*. Other writing that encounters the science performance and its intersection with issues of access and oppression has been published in the journals *PARtake, Global Performance Studies, Comparative Drama*, and elsewhere. She is a former fellow of Fulbright and the Huntington Library and has received funding from NASA's SC Space Consortium and the SC Arts Commission for the development of a multimedia STEAM play about planetary nomenclature. Her critical-creation projects combine puppetry, theatrical devising, and technological play.

Kasi V. Aysola is a dance artist based in the United States. Kasi is a performer, choreographer, and nattuvangam artist whose foundational training in Bharata Natyam was under celebrated Guru Viji Prakash and Mythili Prakash. He had the great fortune to train in Kuchipudi under late Guru Vempati Ravi Shankar and Yamini Saripalli. He cofounded Prakriti Dance, an innovative dance company, in 2014, with the intention to explore the human experience through the idiom of Indian performing traditions. While cultivating and collaborating with artists across America, the company has always focused on empowering artists. He serves as the artistic director of Prakriti Dance, developing and touring original dance works.

Claudia Barnett writes experimental plays about women and history and science, including *Aglaonike's Tiger* and *Kingdom (a play about Snow White and climate change)*. She is a professor of English at Middle Tennessee State University, where she coordinates

In Process: A Creative Writing Event Series and teaches courses in playwriting and dramatic literature, including a new class called Science Plays. Her collaboration with Pipeline-Collective, *Outside of Here*, a twelve-hour play featuring three dozen performers, premiered as a live broadcast event in 2021. Her plays are published by Carnegie Mellon University Press.

Chantal Bilodeau is a playwright whose work focuses on the intersection of science, policy, art, and climate change. Her plays have been presented in a dozen countries, and she is a recipient of the Woodward International Playwriting Prize as well as the First Prize in the Earth Matters on Stage Ecodrama Festival and the Uprising National Playwriting Competition. She is working on a series of eight plays that look at the social and environmental changes taking place in the eight Arctic states. In 2019, she was named one of "8 Trailblazers Who Are Changing the Climate Conversation" by *Audubon Magazine*.

Gianna Bouchard is Associate Lecturer in Theatre at the University of Birmingham, UK. Her research focuses on exploring medical discourse and practices, such as anatomical dissection, in relation to contemporary performance. In *Performing Specimens: Biomedical Display in Contemporary Performance* (2021), she explores the staging and the display of the 'body as specimen' in a variety of performance contexts. She is coeditor, with Alex Mermikides, of *Performance and the Medical Body* (2016) and of two issues of *Performance Research*, "On Medicine" (2014) and "Staging the Wreckage" (2019). Her work has also been published in various edited collections.

Felipe Cervera is Assistant Professor of Theatre and Performance Studies at the University of California, Los Angeles. His research interests are collaborative pedagogies and research methodologies, as well as the interplays of performance, technology and politics. He is Co-editor of Global Performance Studies, and Associate Editor of Performance Research. He serves in the Executive Committee of the Theatre and Performance Research Association (TaPRA) and in the Board of Directors of Performance Studies international (PSi). Publications and artistic work can be found at: www.felipecervera.me.

Meredith Conti is Associate Professor of Theatre at the University at Buffalo, SUNY, USA, and a historian of nineteenth-century performance and popular culture. She is the author of *Playing Sick: Performances of Illness in the Age of Victorian Medicine* (2019) and has articles published in *Theatre Journal*, *Theatre Survey*, and *Studies in Musical Theatre*, among others. She has coedited several volumes: the two-volume collection *Identity, Culture, and the Science Performance*, with Vivian Appler (2022), and *Theatre and the Macabre*, with Kevin Wetmore (2022).

William DeMeritt is a graduate of the Yale School of Drama and an actor, director, narrator, writer, dialect coach, teaching artist, and native New Yorker. An award-winning voice-over artist and narrator, William has been featured in a wide variety of genres and mediums, from commercials to video games to podcasts to audiobooks. As an actor, William has performed in some of the most respected theatres in the country, from the Oregon Shakespeare Festival to Lincoln Center. He can be seen (somewhat regularly) on television, mostly HBO (*The Flight Attendant*, *The Normal Heart*). As an educator he has led workshops, taught classes, and headed up projects for a wide range of educational institutions and businesses, ranging from SUNY Purchase to Apple Inc. @demeritt. www.williamdemeritt.com.

Teófilo Espada-Brignoni holds a PhD in Psychology (Social-Community Psychology Area) from the University of Puerto Rico, Río Piedras Campus. He has taught psychology and social sciences in Puerto Rico and Ohio. He is the author of *The Performance of Authenticity: The Makings of Jazz and the Self in Autobiography* (2022). His research interests include the representation of Puerto Rico during and after the Spanish-American War, recent social justice movements in Puerto Rico, and the life stories and narratives of musicians. Espada-Brignoni is currently Assistant Professor of Psychology at the University of Puerto Rico, Río Piedras Campus.

Mónica I. Feliú-Mójer is a bilingual scientist-turned-communicator who taps into her training, personal background, and culture (a woman from a rural working-class community in Puerto Rico) to engage underserved and overlooked audiences, especially to Puerto

Ricans and other Spanish speakers, with science. She has more than sixteen years of experience in multimedia science communication, community engagement, media relations, science advocacy, and DEI. Dr. Feliú-Mójer serves as Director of Communications and Science Outreach for Ciencia Puerto Rico and as Director of Inclusive Science Communication and Engagement for the Science Communication Lab. She obtained a BS in human biology at the University of Puerto Rico in Bayamón and a PhD in neurobiology at Harvard University.

David Geary (Taranaki Māori) grew up in Aotearoa New Zealand immersed in the Polynesian trickster tales of Maui and is now honored to live in the lands of the Coyote and Raven tricksters of Turtle Island/Canada. He is a playwright, screenwriter, fiction writer, and poet. David teaches screenwriting in the IDF Indigenous Digital Filmmaking program as well as documentary and playwriting at Capilano University, Canada. He wrote three short plays for Climate Change Theatre Action, http://www.climatechangetheatreaction.com/, and his new full-length play *QEIII— Black Betty* (a lost Shakespeare play) will premiere in 2023. He is a member of LMDA Literary Managers and Dramaturgs of the Americas.

Derek Gingrich earned his PhD from York University, where his dissertation examined the cultural uptake of quantum mechanics in German-, French-, and English-speaking countries via science-engaged theatre makers. He is currently Associate Director of Anthropological Research (US) at Lux Research Inc., where he advocates for a human-centric approach to solving the problems of environmental sustainability (renewables, biodegradables, carbon neutralization) and social sustainability (social justice, labor). There, he advances a humanities-focused approach to deciphering the implications of climate change on human lives.

Martine Kei Greene-Rogers is the dean of the Theatre School at DePaul University, USA. As a freelance dramaturg, she has worked at organizations such as the Cleveland Playhouse, the Goodman, the Oregon Shakespeare Festival, Salt Lake Acting Company, Pioneer Memorial Theatre, and the Court Theatre. Her publications include the article "Talkbacks for 'Sensitive Subject Matter' Productions:

The Theory and Practice" in the *Routledge Companion to Dramaturgy* and "A New Noble Kinsmen: The Play On! Project and Making New Plays Out of Old" in *Theatre History Studies*, which she coauthored with Alex N. Vermillion.

Lauren Gunderson has been one of the most produced playwrights in the United States since 2015, topping the list twice including 2022/3. She is a two-time winner of the Steinberg/ATCA New Play Award for *I and You* and *The Book of Will*, the winner of the Lanford Wilson Award, and a finalist for the Susan Smith Blackburn Prize. She is a playwright, screenwriter, musical book writer, and children's author who lives in San Francisco. She graduated from NYU Tisch as a Reynolds Fellow in Social Entrepreneurship. LaurenGunderson.com.

Emily B. Klein is Professor of English at Saint Mary's College of California, USA, where she teaches courses in political theatre and film, performance theory, and gender studies. Her most recent work, "A New Feminist Absurd?: Women's Protest, Fury, and Futility in Contemporary American Theatre," appeared in *Modern Drama* in 2022. She is also coeditor of *Performing Dream Homes: Theater and the Spatial Politics of the Domestic Sphere* (2019) and author of *Sex and War on the American Stage: Lysistrata in Performance 1930–2012* (2014), which was featured in *The New York Times*, *Ms.*, and *Vice*.

Kirsten Lindquist is of Cree-Métis and settler ancestry from rural northeast Alberta and is currently a PhD student in Indigenous Studies at the Faculty of Native Studies, University of Alberta, Canada. She introduced and taught a new selected topics course to the faculty, Indigenous New Media, which integrates both classroom and lab learning experiences. In her role as co-producer for Tipi Confessions, she is also a research assistant for Relab, a research-creation laboratory.

Alexis Riley is a disability performance scholar, artist, and educator whose research focuses on the politics of mad and disabled embodiment. Her recent publications have appeared in *The International Review of Qualitative Research* and *Theatre Topics*. Alexis is currently a President's Postdoctoral Fellow at the University of Michigan, where she is working on a book project that explores performances enacted in state hospitals as an embodied record of mad-crip cultural memory.

Diane Stubbings is a playwright and critic. Her work has been performed in Australia and New Zealand and shortlisted for awards including the Patrick White Playwrights' Award, the Rodney Seaborn Playwrights' Award, and the Griffin Award. Diane recently completed a PhD at the Victorian College of the Arts, University of Melbourne, where she used practice-based research to investigate intersections between science and theatre.

Kim TallBear (Sisseton-Wahpeton Oyate) is Professor and Canada Research Chair in Indigenous Peoples, Technoscience, and Society, Faculty of Native Studies, University of Alberta, Canada. She is the author of *Native American DNA: Tribal Belonging and the False Promise of Genetic Science*. In addition to studying genome science disruptions to Indigenous self-definitions, Dr. TallBear studies colonial disruptions to Indigenous sexual relations. She is a regular panelist on the weekly podcast, Media Indigena. You can follow her research group at https://indigenoussts.com/. She tweets @ KimTallBear. You can also follow her Substack newsletter, *Unsettle: Indigenous affairs, cultural politics & (de)colonization*.

Mike Vanden Heuvel is Professor of Interdisciplinary Theatre Studies and a member of the Department of Classics and Near Eastern Studies at the University of Wisconsin-Madison, USA. He recently edited the two-volume collection for Bloomsbury, *American Theatre Ensembles*. Recent publications include work on devised theatre and science and the intellectual traditions behind Tom Stoppard's plays.

Madhvi J. Venkatesh is an educator, dancer, and scholar who develops and studies interdisciplinary learning environments that build students' professional competencies, self-efficacy, sense of belonging, and metacognitive skills. She is an Assistant Professor of Biochemistry and the Curriculum Director of the Interdisciplinary Graduate Program in Biological and Biomedical Sciences at Vanderbilt University School of Medicine, USA. Madhvi is also a professional Bharata Natyam dancer and received her training in the art form from Viji Prakash. She performs across the United States and India as a soloist and member of Prakriti Dance, an innovative US-based company which she cofounded in 2014.

INDEX

4.48 Psychosis (2000)   48

ableism   39, 50
Abuzz   126, 127
academia   15, 75, 120, 125–6, 130–2
   in Germany   180–7
acoustics   9 n.4, 113–14, *see also* music; noise; sound
acting   44–7, 49, 53 n.19, 110
activism
   climate   22, 32 n.7
   Hawai'ian   141–2
   women   126
activist(s)   124
   mad   38
Adelphi Theatre   214
aesthetics   41–2, 49, 51, 62, 112
Agricultural Experiment Station (AES)   164, 165, 167–9, 171–3, 176 n.42, *see also* *Estación Agronómica*
agricultural science   7, 161–2, 164–72
Alaska   14, 15, 17, 136
Alexander, Stephon   113
Alfred P. Sloan Foundation   104, 204, 218 n.17
Alhous, Peter   128–9
Alpha Generative Engine   112
American Psychiatric Association (APA)   41, 43
Anchorage, Alaska   14

animal(s)   4, 17–18, 22, 29–30, 55–7, 65–9, 125, *see also* more-than-human; nonhumans
   divide between humans and   57–8, 96
   emotions   56–7
   industry   171
animation   29, 110, 121, 128, 130–3
anti-science movement   7, 186, 190–203, 214–16
anti-Semitism   180, 186, 190–1, 193
Aotearoa   27, 28
apocalypse   20, 148
Appler, Vivian   13, 138, 143, 228
Archer, Samantha   19–20
art   37, 181, 203
   anti-Nazi   193
   conceptual   111
   digital   111–12
   installation   111
   performance   76
   public   193, 215
   and science   3–4, 6, 62, 64, 102–6, 108, 115, 130, 211–12, 214, 240
artist   19, 124–5, 228
   Indigenous   15
   multimedia   102
   performance   111, 144
   remix   112, 114

INDEX

Arts & Climate Initiative   14
Assisted Reproductive Technology (ART)   83–5, 89
asthma   81, 122, 124, 129
astrology   2, 9 n.5, 9 n.6, 140, 191
astronomy   6–7, 102, 108–9, 139–41, 143–4, 189, 191
astrophysics   109–10, 140
audience engagement   115, 201
Austin, Texas   23
Avendaño, Lukas   7, 144–9

Baal (1923)   182, 195 n.18
Background to Breakthrough (B2B)   121, 127–8, 131–4
baking   16, 34
Barber, Philip   209–10
Barnett, Claudia   8, 13, 93
Bay Area Science Festival   133
Bear, Tracy   14
Beckett, Samuel   8, 222
Berkeley Repertory Theatre   201
Bharata Natyam   5, 58, 61–2, 67–8, 71 n.10
Big Hand for Science, A (2019)   211–12, 215
Bilodeau, Chantal   13, 153
biocapitalism   6, 77, 83, 86–7
bioengineering   126
biology   8, 56–8, 76, 80, 85–6, 89, 94–8, 123–5, 186, 197 n.48, 212, 222–3
biomedicalization   6, 76–7
biomedical science   6, 81, 121–3, 128, 130–1
Birth of a Nation (1915)   103
Birth of Tragedy   184
Biscuit, Rebecca   77–82, *see also* Sh!t Theatre
Blue Planet II (2017)   59
blues   103, 111–12
bodies   21–3, 39, 46, 76, 125

actors'   40, 48
dancing   59, 62, 67, 69
female   69, 81, 83, 85, 87–90
gendered   145
Indigenous   145–7
and labor   6, 76, 85, 88
language   58
mad   42
materiality of   76–7
as metaphor   164
othered   17, 90
parts   86
performing   39, 49
racialized   90
sovereignty   23
surrogate's   86
women's   81, 83–5, 88–9
Bodies (2017)   77, 83–5, 87–9
Bohr, Niels   180, 185, 188–90, 197 n.47, 197 n.48
Born, Max   185, 197 n.48
Boston, Massachusetts   23, 70, 113
Brantley, Ben   46
Brecht, Bertolt   7, 179–84, 186–94
Bristol Old Vic   1–3
Brother Buffalo   28–9, 33 n.15
Burchard, Esteban González   122–4, 128–33
Butler, Judith   39, 51 n.4, 161–2, 171

Calisi Rodríguez, Rebecca   122, 124–6, 128–30
Calouste Gulbenkian Foundation   104
Canada   14, 20, 34–5, 142
Carnatic music   59, 71 n.10, 72 n.15
Casimir, Hendrik B. G.   186
catastrophe   229–31

Catastrophist, The (2021)   7, 8, 231–46
Catholicism   147–8, 181
causality   184–5, 190–1
celestial politics   138–9, 141, 146–9
"Charting an Original Path"   124–6, 129
Chaudhuri, Una   65, 68
chemistry   18–19, 59, 68, 126, 202, 222
Chicago Tribune   37
children's literature   163
class   69, 83, 85, 86, 121, 123, 140–3, 180–2, 208
classism   4, 87–8, 121, 123, 140, 208
climate   5
  activism   22, 32 n.7
  breakdown   22
  change   8, 15, 17–20, 25, 28, 104, 111
  change anxiety   8
  crisis   17–18, 24, 97–8
  disaster   201
  justice   22
Climate Change Theatre Action   14, 24
clinic   41–2, 49–50
clinical trial   4, 77–82, 90
collaboration   22, 26, 76, 80, 101–3, 105, 108, 111–12, 115, 213–15, 231–4, 242
colonial/colonialism   5, 14, 17, 24, 27–8, 87, 103, 139–45, 147–8, 160–1, 163–4, 172–3
  science   145–6, 172
  settler   141
  tradition   8
colonization   18, 148, 160, 168–9
  performance of   165
Columbus, Christopher   35, 206–7, 209

communication   62, 171
  channel   107–8
  of science   104, 115, 121, 133–4
Complicité   75
Conti, Meredith   13, 52 n.6, 138, 228
Copernicus, [Nicolaus]   140
coral reefs   5, 62–3, 65
cosmic
  agency   142
  elite   139–40, 142–3
  underclass   140–3
Costa Rica   65
costume design   59, 62–3
Covid-19   7, 126, 201–3, 210, 214–15, 228–31, *see also* pandemic
Cree   25, 31 n.4
critical mad studies   38, *see also* madness
critical outer space studies   143–4, *see also* cosmic; outer space
Critical Polyamorist 100s   24–5, 34–5
culture   5, 62, 112, 184, 231
  American   201, 216
  European   182
  Indigenous American   33 n.15
  mad   51
  science as   162–3
  Western   180–2, 184
curlew   145

Dallas Zoo   124–5
dance   61–3, 65–6, 69, 77
Decline of the West   180–1
decolonial/decolonization   3, 14–15, 22, 142, 148–9
  and genomics   14
  and Indigeneity   15
deficit narratives   121
DeMeritt, William   228

Democratic Republic of
    Congo   125
Denno, Lydia   88
de Olmos, Andrés   148
design   48–9, 59
    costume   59, 62–3
    lighting   47, 62–3, 109
    projection   47, 49, 102,
        109, 113
    sound   1–3, 9 n.3, 47, 59,
        109, 110, 130–1
*Diagnostic and Statistical Manual
    of Mental Disorders, The*
    (*DSM*)   41–3
Disappearing Number, A
    (2007)   75
diversity   4, 6, 121, 132, 214
    biological   64
    representation of   3
    in science   120–2, 133–4
diversity, equity, and inclusion
    (DEI)   121, 131, 134
DJ Spooky   102–3, 111–14
DNA   19, 75, 95, 222–3
documentary
    film   6, 15, 121, 129–30
    theatre   202, 211, 217 n.7
Donald, Dwayne   25
dramaturgy   40, 48, 222–3,
    234–5, 237, 241, 243 n.9
    biological   8, 223
    representational   114

early modern   1, 3, 9 n.5, 9 n.6,
    76
Earth   8, 24, 27, 58, 96, 140,
    188–9
Edmonton, Canada   14, 22–3,
    27, 34
education   3, 7, 23, 189, 215,
    236–7
    disparities in   121
    K-12   4

medical   211
public   215
in science   126
Einstein, Albert   184–6, 190–2,
    231
embodiment   5, 38–40, 42, 45,
    49, 60–2, 64, 67, 76–7, 81,
    83, 88–9, 144, 160, 234–6
emotion   4, 9 n.5, 21, 55–7,
    59–62, 69, 130–1, 133, 235
empathy   5, 57–9, 61, 69, 108,
    111, 128, 130, 214
environment   14, 17–19, 23–4,
    26–7, 65, 125, 223, *see also*
    climate
    cellular   222
    conservation of   69, 96
    crisis   111
    destruction of   21, 59, 69
    intellectual   181–2
    protection of   67
Epic theatre   183, 188
Estación Agronómica   166–7,
    see also Agricultural
    Experiment Station
experiment   66, 75–7, 79, 81–2,
    114–15, 126, 169, 171–3,
    188–9, 205, 227, 240
    artistic   66–7, 121, 130, 204,
        211
    on bodies   80–2
    dramatic   188
    public   114–15, 171
    tools   188–9
    Tuskegee Experiment   208
extractivism   17, 21, 28, 32 n.7,
    32 n.12, 65, 69, 155

Fairchild, David   169
fascism   180, 186–7, 190, 200–1
Fauci, Anthony   230
*Fear and Misery of the Third
    Reich* (1938)   192–3

Federal Theatre Project
   (FTP)   200–11, 215–16
Figgener, Christine   65
film   6, 27, 29, 39, 49, 52 n.5, 77, 103, 120–2, 127–34, 233, 239, 242
Final Judgement, The
   (c. 1531)   148
"Finding Sublime in the Mundane"   126–7
Finkle, David   46
First World War   180, 181, 183
Flanagan, Hallie   204, 208–9, 214
Florida   58, 169
Flüchtlingsgespräche   187–8
Foldscope   126–8, 130
Franklin, Rosalind   75
Franzmann, Vivienne   77, 84, 87, 89
frugal science   126–8

Gaiger, Jason   183–4
Gale, Elbridge   169
[Galilei], Galileo   140, 181, 188–94
gardens   27–30
Gardner, Frank D.   167–8, 171
gay   19, 43, 103, *see also* LGBTQ+
Geary, David   13
gender   18, 20–1, 28, 120, 144–8, 232
   binary   21, 148
   body   145
   discrimination   18, 238
   fluidity   144
   hierarchies   68–9
   identity   68, 109, 144
   nonconforming   43, 47, 120
   performativity   43, 47
   in science   120–1, 149, 239
   studies   19–20

and violence   21
Genealogy of Morality   184
George, Marian   164
Georgian [era]   1–3, 9 n.3, 9 n.4
Germany   181–2, 192
   Nazi   179, 197 n.47
gestus   190–1
Ghostkeeper, Elmer   25
Glaspell, Susan   206
Gleiser, Marcelo   113
Globe Theatre   2, 10 n.7
Goethe, Johann Wolfgang von   180, 184, 186
Goffman, Erving   161–2, 171
Going Dark (2012)   102, 108–10, 113–14
Gore, Al   15
Grand Canyon Park   30
Great Depression   201, 203–4, 208
Greene, Jesse   242
Green-Rogers, Martine Kei   228
Griffin, Nona (Nina)   123, 128
Griffith, D. W.   103
Guelaguetza Festival   145–6
Guinea Pigs on Trial
   (2014–16)   77–82
Gunderson, Lauren   8, 228, 244

hacking   103, 112–13
Handbook of Physics   185, 188
Hand to Hand (2019)   204–5, 211–13, 215
Harding, Sandra   142–3, 149, 162–3
Harr, Larry   205–6, 215
Harvard University   124, 245
Hatch Act   165–6
Hauser, Ethel Aaron   209–10
Hawai'i   141–2
Hayden Planetarium   113
Heinz Endowments   213

INDEX

Heisenberg, Werner   180, 185–7, 192–3, 197 n.47, 197 n.53
Henning, Joel   46
historiography   7, 39–40, 147, 193
   mad   38–9
   and performance criticism   40–1, 45, 50–1
   of science   141
Hitler, Adolf   179, 186, 201
homosexuality   43, 146, *see also* LGBTQ+
horticulture   169–70
House Committee on Agriculture   167–8
human(s)   9 n.6, 66–7, 95–6
   action   9 n.5, 65
   behavior   2, 59
   divide between animals and   57–8, 96
   emotions   57
   existence   55
   experience   3–4, 18, 71 n.9
   identity   4
   life   222, 227, 233, 240
   and nature   24, 57–9, 68
   and nonhumans   5, 17, 76
   sexuality   15, 24–5
   superiority   58–9, 69
hummingbirds   29
Hunter, Lynette   139

identity
   African American   123–4, 129–30, 132, 208
   American   168, 202, 212
   Black   20, 120, 122–3, 238
   disabled   51, 120
   gender   68, 109, 144
   German   181–3
   human   4
   Indigenous   34–5, 144
   Jewish   238–9, 244–6
   Latine   123, 129
   marginalized   120–1, 126
   Mexican   144–5
   Mexican American   123, 124, 130
   mixed-race   123
   personal   8
   Puerto Rican   160–1
   racial   123, 132
   and science   122, 125
   social   8, 147
Identity, Culture, and the Science Performance, Volume 1   4, 138
imperialism   142–3, 160–8, 170–3, *see also* colonial/colonialism
"Inclusive Future, An"   123–4, 132
Inconvenient Truth, An (2006)   15
Indecent (2019)   238
India   59, 61, 67, 83, 126–8, 168
Indian Act   25–6, 32 n.11
Indiana Jones   237, 239
Indigenous   20, 22–3, 29–30, 144–8, *see also* Aotearoa; Cree; Métis Nation; *muxe*; Sisseton-Wahpeton Oyate; Stó:lō Nation; Taranaki Nation; Zapotec
   artists   15
   bodies   145–7
   cultures   6, 23, 33 n.15
   and decolonization   15
   feminism   26
   identity   34–5, 144
   languages   15
   scholars   15
   tradition   6–7, 21, 32, 145
   women   14, 17, 23, 145, 148
informed consent   79–80

interdisciplinarity 3–5, 37–8, 42, 102–6, 108, 110–12, 114–15, 143
Isherwood, Charles 45–6
It Can't Happen Here (1936) 200

Jackson-Schebetta, Lisa 161, 168
James, Clive 222
jazz 108, 111–13
Jim Henson Foundation 213
journalism 37, 162–3, 202–3
Judaism 237–9, 244–6
Jurassic Park (1993) 237, 239

Kahlo, Frida 145
Kaiser/Silhouette, The 181–2
Kane, Sarah 48
Kant, Immanuel 180–2, 185–6
Kaur Chohan, Satinder 83, 85–7
Keller, Julia 37, 49–50
Kepler, Johannes 110–11, 140
Kepler Story, The (2013) 102, 110–11, 113–14
Kingdom (a play about Snow White and climate change) 18
Kinman, Charles Franklin 169–70
Kinney, Terry 44, 46
Kirschner, Elliot 128, 133
Knapp, Seaman Asahel 165–6, 168, 170–1
Koenig, Rhoda 47
Krapp's Last Tape (1958) 8, 222
Krisis der Wissenschaft 180, 183–4

labor 140, 146, 167, 204
    affective 84, 87
    biological 80, 85–6
    and bodies 6, 85, 88
    clinical 6, 78, 82–3, 85–8, 90
    emotional 85–6
    feminized 88, 90
    intellectual 186
    precarious 77–8, 84, 90, 203
    reproductive 6, 83, 89
    scientific 75–6, 88–9
    theatrical 204–5
laboratory 6, 19–20, 75–6, 83, 95, 125, 126, 208
Ladies Garment Workers Union 210
La Fiesta del Árbol 170
land 15, 17, 20, 22, 24, 32 n.7, 139, 141–8, 160, 166, 186
Lascassas, Tennessee 13
Lebensphilosophie 180–1, 183–6, 189, 192
Lenard, Philip 186
Lewis, Sinclair 200–1
LGBTQ+ 120
Life of Galileo (Leben des Galilei, 1938) 7, 179–81, 187–94
lighting design 47, 62–3, 109
Lindquist, Kirsten 13, 16
Living Newspapers 201–3, 206–7, 210, 214–16

McKay, Carman 29–30, 33 n.16
McMaster University 13
Made in India (2017) 77, 83, 85–90
madness 3, 48–9
    definition of 38
    historiography of 38–9
    perceptions of 44
    performativity of 39–40, 43, 45, 49–51
    representations of 45–6, 50
    and scientists 67, 120
Magelssen, Scott 143
Magli, Giuglio 139, 140
mangoes 159, 161, 168–70
Mann ist Mann 187

Māori   28, 33 n.12
March for Science   130, 137 n.40
marine life   4, 5, 55–60, 63, 65–6, 68–9
Marin Theatre   228, 233, 234, 243
Marxism   180, 182–4, 186, 188–9, 191–2, 194
mathematics   19, 75, 184, 190–1, 195 n.9
Mauna Kea   141–2
media   131
    digital   119
    newspapers   119, 159
    social   65, 119, 126, 132, 216
medicine   41, 43, 76, 122–4, 208, 230, *see also* public health
    racism in   128–31
mental health professionals   40, 42, 50, 52 n.5
mental illness   37–8
method acting   45, 52 n.17, 53 n.19
Me-Ti: Buch der Wendungen   191–2
Métis Nation   14
Metropolitan Theatre   209
Mexico   144–8
microscope   126–8, 211–12
Mignolo, Walter   160, 161
Miller, Paul D., *see* DJ Spooky
Minadakis, Jasson   233, 234
Minneapolis, Minnesota   13, 35
Møller, Christian   189
Montreal, Canada   14
More Good Sports   186
more-than-human   15, 17, 21, 33 n.16, *see also* nonhumans
Mother Girth   20–1
motherhood   34, 84–5, 97, 126–7, 153–6, 224
Mothersole, Louise   77–82, *see also* Sh!t Theatre

music   34, 47, 110–14, 131, 212, *see also* blues; jazz
    Carnatic   59, 71 n.10, 72 n.15
    powwow   21
Mussolini, Benito   201
muxe   6–7, 144–7

Nahuas   147–8
narrative   8, 17, 20, 58–9, 62, 109, 121, 130–1, 143, *see also* storytelling
National Academies Keck *Futures Initiative* (NAKFI)   64, 69, 73 n.28
National Academy of Science   133
National Endowment for Science, Technology, and the Arts   104
National Endowment for the Arts   213, 218 n.16
National Geographic   65
National Research Mentoring Network   132
National Venereal Disease Control Act   206
Naylor, Hattie   102, 108
Nazism   184–7, 192–4, 197 n.47
neoliberalism   103–4
Neukom Institute of Computational Science   103, 111–12
neurobiology   43, 46, 124
neuroscience   48, 113
New York City   14, 132, 209
New York Times   46, 242
Next to Normal (2008)   48–9
Nielsen, Anthony   48
Nietzsche contra Wagner   184
Nietzsche, Friedrich   180, 182–4, 196 n.32
noise   106–8, 110–11, 115, 225–6

No More Harveys (2022)   21–2, 32 n.6
nonhumans   13, 15, 17–18, 29, 56–7, 62–3, *see also* more-than-human
and humans   5, 17, 76
Northwest Passage   105–6
Norway   18, 26
nuclear weapons   19

Oaxaca, Mexico   144–6
Ober, Frederick A.   163–4, 172
observation   41, 62, 64, 120, 190, 192, 211
ocean   5–6, 22, 55, 59, 62–6, 69, 93–8, 106, 153, 226
   awareness   57, 64, 73 n.28
   representations of   63–4, 68
O'Hearn, Steve   212–14, 216
okapi   125
One Flew Over the Cuckoo's Nest
   on Broadway (1963)   39, 40
   film (1975)   39, 49
   novel   39, 43–4
   prequel to   39
   Steppenwolf production (2000)   36–7, 40, 43–5, 48–50
One-Third of a Nation (1938)   206, 214
ontology   3, 105, 108, 114–15, 143, 149, 160, 163
Oregon Social Hygiene Association   210
Oregon State Medical Association   210
O'Reilly, Kira   76
Orientalism   160
ORLAN   76
othering   17–18, 48, 160
otherness   17–18, 155
outer space   139–40, 143–4, 149, *see also* astronomy; astrophysics; cosmic

Overfield, Richard A.   165

pandemic   7, 17, 203–5, 210, 213–17, 228–31, 233–5, 237, 242, *see also* Covid-19
Paperfuge   126
parasite   107–10, 112, 114–15
Parasite, The   107–8
Parran, Thomas   206
pathology   42, 109
Pennsylvania Council on the Arts   213
performance   5, 38–9, 41
   art   76
   of colonization   165
   criticism   39, 42, 49–50
   devised   75, 77, 111
   immersive   109–10, 113
   institutional   167
   political   139, 148–9
   practices   211
   public   204
   and science   3–4, 7, 41–2, 49, 76, 80, 115, 162, 171, 204–5, 211, 236
   science-oriented   18, 204
   textual   160, 163–4, 167, 173
performance studies   39, 139, 143–4
performativity   24, 39, 40, 143, 145, 161–2
   gender   43, 47
   of imperialism   162, 164–7, 170–1
   mad   5, 39–45, 49–51
personhood   28–30, 32 n.12
Pew Research Center   119
Pez Globo (Pufferfish)   126
Phlebotomist, The (2018)   76
Photograph 51 (2008)   75
physics   10 n.12, 113–14, 180–7, 189–91, 193
pigeons   125, 129

Pittsburgh, Pennsylvania   7, 211, 213–14
Pittsburgh Foundation   213
Planck, Max   183, 195 n.12
planetarium   110, 114
  lecture   109, 113–14
  show   102–3, 105, 108–9, 111, 114
plastic   59, 68
  pollution   5–6, 59–60, 62, 65–8
plays   14–18, 148, *see also* dramaturgy; playwriting; theatre
  audio   233
  Broadway   40
  Catholic   148
  development   206
  one-person   232–3
  public health   204
  science-integrative   3
  and scientists   240
playwriting   15, 17–18, 24–6, 28, 222–3, 232–3, 239–41
practice as research   121, 133
prairie   22, 25
Prakash, Manu   122, 126–8, 130–1, 133–4
Prakriti Dance   57, 70, 71 n.9
Prieto, Antonio   145, 146
Principles of Atomic Physics   187
projection design   47, 49, 102, 109, 113
psychiatry   36–8, 40–9, 52 n.5
Ptolemaic system   188, 190
Public Enemy   103
public health   122, 204–6, 208, 211, 214–15, 236–8
Puerto Rico   7, 121, 159–73, 176 n.42

QTPOC   20, *see also* LGBTQ+
quantum

computation   113
  mechanics   184–90, 192, 197 n.48
  physics   10 n.12
  "River"   34–5
queer   20–1, 77, *see also* LGBTQ+; QTPOC; two-spirit LGBTQIA+
Quintanilla, Guillermo   166–7

race   102–3, 186, 192, 201
  and discrimination   133, 238–9
  and epistemology   143
  and identity   123, 132
  and medicine   123, 130–3
  and privilege   85, 88
  and science   119–20, 122, 231
  and stereotypes   48, 129
  and surrogacy   82, 84–8, 90
racism   22, 103, 122–3, 131, 142
  medical   122–4, 132–3
raga   59, 72 n.15
Rampur, India   126–7
Rancière, Jacques   141
realism   44–5
Rebirth of a Nation (2007)   103
Relab   14, 20, 31 n.1
remix   6, 103, 107–8, 112–15
reproductive labor   77, 83–9
reproductive tourism   87
Requiem para un Alcaraván (2015)   145–9
rhythm science   103, 112, 114
"River Quantum"   34–5
Rjukanfossen   18
Road, Ella   76
Royal Court Theatre   48, 83
Ruocco, Peter   233, 234

Said, Edward   160–1
Saint Paul, Minnesota   13

San Francisco, California 23, 122, 123
Schopenhauer, Arthur 182–4
Schreck, Heidi 232
Schrödinger, Erwin 185
sciart 104, 115
science, *see also* specific disciplines
 and art 3–4, 6, 62, 64, 102–6, 108, 115, 130, 211–12, 214, 240
 authority of 102, 113, 190
 careers 121
 colonial 145–6, 172
 communication of 104, 115, 121, 133–4
 and ethics 76–7
 filmmaking 121
 frugal 126–8
 and gender 120–1, 149, 239
 journalism 37
 labor of 75–6, 88–9
 and performance 3–4, 7, 41–2, 49, 76, 80, 115, 162, 171, 204–5, 211, 236
 postcolonial 145–6
 precolonial 145–6
 and race 119–20, 122, 231
 rhythm 103, 112, 114
 Western 162–3, 180–1, 184
Science Communication Lab (SCL) 120–1, 134
science performance 4–7, 102–3, 111, 138, 140–1, 143
 definition 4–5
Scientific Revolution 110–11
scientist(s) 55, 65, 97, 113, 119–21, 167, 178–80, 186, 194, 209, 214–15
 of color 120–1, 131, 134
 and gender 121
 identity 122
 mad 67, 120
 marginalized 121, 122, 126, 131, 134
 minoritized 120–1, 131
 and parenthood 125–6
 representations of 75–6, 120, 127–9, 229, 236–41
 Western 162
Seattle, Washington 19, 209, 210
sea turtles 62, 65–6, 68, *see also* marine life
Second World War 179, 181, 185
sense 244, *see also* smell
Serres, Michel 101, 105–9
"Sex at the End of the World" (2018) 20
sexism 4, 39, 232, 239, 242
sexuality 14–15, 20, 43
Sh!t Theatre 77, 81–2
Shakespeare [William] 2, 231
Shepherd-Barr, Kirsten 52 n.6
Sisseton-Wahpeton Oyate 13
sky 6, 139–42, 144, 146–9, *see also* outer space
smell 159, 235–6, 244–5
Smiths, The 33 n.17
Snow, C. P. 102
social change 57, 63–5
social justice 69, 88, 186, 246
social norms 43–4, 144–5
Society for Advancement of Chicanos, Hispanics, and Native Americans in Science 133
Society for Neuroscience 126
Sofer, Andrew 3–4
So Spoke Zarathustra 182
sound design 1–3, 9 n.3, 47, 59, 110, 130–1
Sound&Fury 102, 108–9
South Dakota 13
Spain 147, 159–60, 166–7, 172

Spanish-American War   7, 159–61, 165–6, 172–3
Spengler, Oswald   180, 184–7
Spirochete (1928)   204–10, 215
Spivak, Gayatri   141
Sputnik   102
Squonk   7, 204–5, 211–16
Squonk in the Neighborhood (2021)   211–14
Stalin, Josef   179, 186
Stanford University   126, 131
Stark, Johannes   186, 192
static   107–8, *see also* noise
STEAM   204, 211–12
STEM   3, 6, 119
Steppenwolf Theatre Company   5, 36–8, 48, 50, 53 n.17
Stó:lō Nation   29
storytelling   3, 14, 23, 52 n.5, 58, 65–6, 112, 121, 127–9, 133, 135 n.13, 242, *see also* narrative
Sturm und Drang   185
Subliminal Kid, That, *see* DJ Spooky
Summer internship for INdigenous peoples in Genomics (SING)   14, 19
Sundgaard, Arnold   206
surrogacy   6, 77, 82–90
SymbioticA   76
syphilis   206–10, 215–16, 220 n.33

Tagaq, Tanya   28, 33 n.14
TallBear, Kim   8, 13, 16, 34
Tamasha Theatre Company   85
Taoism   180–1, 183, 191, 192
Taranaki Nation   27
Taylor, Diana   143
technology   3–4, 9 n.6, 76, 160, 170, 181–2, 201–2
studies   142
theatre   1–2, 9 n.4
TED Talk   128, 235
Tehuantepec Isthmus   145, 148
telescope   141–2, 190, 211
Tennessee   13, 26
Tennessee State University   13
textual performance   160, 163–4, 167, 173
theatre   24, 29, 39, 50, 52 n.6, 76–7, 81–2, 139, 148, 182–3, 201–10, 215–16, 234–6, 241–2
  architecture   1–3
  contemporary   77
  criticism   5, 37–8, 49–51, 242
  documentary   211, 217 n.7
  information   211
  and labor   204–5
  and science   3–4, 7, 101–5, 108, 110–11, 211, 231–3, 240
  technology   1–2, 9 n.4
Theatre des wissenschaftlichen Zeitalters   179, 193–4
theatricality   4, 42–3, 102, 110–11
thermal exciter   106–7, 113
Third Reich   186, 190–2, *see also* Nazism
Thirty-Meter-Wide Telescope (TMT)   141–2
Thompson, Lanny   163
Through Fish Eyes (2019)   5, 6, 57–64, 66–9
thunder run   1–4, 8 n.2, 9 n.3, 9 n.4, *see also* sound design
Tipi Confessions   14–15, 17, 19, 23, 25–6
Tolleson, Jeff   202–3
Toronto Queer Film Festival   17, 22
tradition

blues   103, 112
colonial   8
cultural   6–7
dance   21, 57–8, 71 n.10, 145
dramatic   182
epistemological   142–3
Indian   57–8, 71 n.10
Indigenous   6–7, 21, 32, 145
land   32
production   24
scientific   75–6, 113, 162–3, 185, 190
Tragedy of King Lear, The (1605-8)   1–4
trauma   43, 44, 46–7, 65
Treaty of Paris   160
True, A. C.   167, 171
Trump, Donald J.   201–5, 212, 214, 218 n.16
Turing, Alan   103
Tuskegee Experiment   208
two-spirit LGBTQIA+   17, 20, 23, *see also* LGBTQ+

uncertainty principle   187–8, 194
United States Department of Agriculture (USDA)   164–5, 167–9, 173
United States Supreme Court   160
University of Alberta   13, 14
University of California Davis   124, 137 n.46
University of Connecticut   19
University of Texas   19
University of Washington   19

vaccine   230–3
  anti-   203
Vancouver, Canada   20
Victoria, Canada   25
virology   7, 8, 232, 245
Visiting Nurses' Association   210

Vogel, Paula   238
von Laue, Max   183, 192

Wagner, Richard   184, 196 n.32
Waldby, Catherine   76, 77, 89
Washington Post   202
Wasserman, Dale   39, 43–54
"Water Spirit's Retribution" (2019)   20–1
Weigel, Helene   193
Weimar Republic   180–4, 190, 192
Wein, Wilhelm   180, 185
Weiss, Heidi   45
Wellcome Trust   104
whales   59–61, 68, 94–8, *see also* marine life
Whanganui River   28
What the Constitution Means to Me (2017)   232
"Which Box Do I Check?"   123, 130–1, 133
Whiskeyjack, Lana   17, 25, 31 n.4
white
  epistemology   143
  people   122–3
  performers   77
  privilege   85, 88
  skin   84
  supremacy   143
  women   83, 87
whiteness   4, 102, 123, 143, 231
"Who's Included?"   122–3, 129
Winnipeg Treaty One   20
Wise, Nina   102, 110–11
Wolfe, Nathan   8, 228–32, 234, 235, 237–41, 244
women   18, 21–2, 68, 89, 95, 153–6, 232
  bodies   77, 81, 83–5, 88–9
  brown   83, 88
  and class   85–7, 121

of color   84, 88
  Indian   86
  Indigenous   14, 17, 23, 145, 148
  racialized   85
  and reproduction   77, 83–90
  in science   120–1, 125–6, 129
  white   83, 87
Wonder Collaborative   6, 120–2
Wonderful World of Dissocia, The (2007)   48
Woolf, Virginia   222–3

Works Progress Administration (WPA)   201–2, 216
Wrangham, Richard   23, 243 n.5
Wu Wei   181, 183, 186, 192–3

X-Files, The (1993–2018)   78, 91 n.7

Zapotec   144–7
Ziegler, Anna   75
Zoom   22, 233–4, 237

www.ingramcontent.com/pod-product-compliance
Lightning Source LLC
Chambersburg PA
CBHW071810300426
44116CB00009B/1260